92 **Topics in Current Chemistry**

Fortschritte der Chemischen Forschung

W0106501

Organic Chemistry

Springer-Verlag
Berlin Heidelberg GmbH 1980

This series presents critical reviews of the present position and future trends in modern chemical research. It is addressed to all research and industrial chemists who wish to keep abreast of advances in their subject.

As a rule, contributions are specially commissioned. The editors and publishers will, however, always be pleased to receive suggestions and supplementary information. Papers are accepted for "Topics in Current Chemistry" in English.

ISBN 978-3-662-15402-1   ISBN 978-3-540-38249-2 (eBook)
DOI 10.1007/978-3-540-38249-2

© by Springer-Verlag Berlin Heidelberg 1980
Originally published by Springer-Verlag Berlin Heidelberg New York in 1980
Softcover reprint of the hardcover 1st edition 1980

2152/3140-543210

Contents

Two Step Reversible Redox Systems of the Weitz Type

Siegfried Hünig and Horst Berneth

Institut für Organische Chemie der Universität Würzburg, D-8700 Würzburg

In Memory of Hans L. Meerwein (May 20, 1879 – October 24, 1965), the pioneer of carbenium ion chemistry, dedicated to his 100th birthday.

Table of Contents

1 Introduction

A great variety of seemingly unrelated organic compounds have been demonstrated to transfer two electrons in a stepwise fashion, if they can be derived from the general structural types A, B or C. The intermediate oxidation level SEM thereby represents radical cations, radical anions or neutral radicals[1a]. Their thermodynamic stability can be understood within a general theory of polymethines $X-(CH)_{N-2}-X'$ containing $N \pm 1$ π-electrones[2a] for which MO-LCAO calculations have been developed[2b].

This review concentrates on a special type which only recently has been investigated thoroughly and which has been designated as *"Weitz* type". Here the heteroatoms X and Y are members of cyclic π-systems showing "quinoid" (or polyenic) character in the reduced form RED. In the oxidized form OX, however, they exhibit "aromatic" behaviour. The potentials E_1 and E_2 can be determined by different voltammetric methods.

The report will emphasize mainly type *A* with a radical cation SEM ("violene"[1b]) as the intermediate oxidation level[3]. Structural types *B* and *C* will be dealt with essentially for the sake of completeness and for comparison.

The investigations reported here have been selected to evaluate general rules by systematic variations with regard to the following properties:

1) The positions of the potentials E_1 and E_2.
2) The difference of E_1 and E_2 measuring the thermodynamic stability of SEM, mainly in systems of type *A*. According to *Michaelis*[4a] the semiquinone formation constant $K_{SEM} = [SEM]^2/([RED][OX])$ can be calculated by means of $E_2 - E_1 = 0.059 \lg K_{SEM}$ (V, 25 °C).
3) UV/VIS spectra of all three oxidation levels, especially SEM, which always shows the longest wavelength absorption.
4) The ESR spectra of SEM from which spin densities may be derived.

3

2 N,N'-Disubstituted Bipyridyls and Dipyridyl Ethenes

2.1 Variation of N-Substituents in 4,4'-Bipyridyl

The redox system 1 (R = alkyl; 1_{OX} = "Viologenes"[4b, 5]) was the first to be interpreted correctly (*E. Weitz*[5]). It is especially well suited for studying the effects of N-substituents because steric effects are virtually absent. In spite of the great importance of some of these quaternary salts as universal herbicides (R = CH_3, "paraquat"®) only potentials E_2 were known for a long period of time, since the reductions SEM/RED in aqueous medium are mostly irreversible[6]. In aprotic media, however, as in acetonitrile and DMF, E_2 and E_1 are ideally reversible[7]. This holds true for most of the investigated substituents as can be seen from Table 1.

$$\underset{1}{\overset{1}{=}}RED \quad \overset{*}{\underset{1}{\rho}}=0{,}256 \qquad \underset{}{\overset{1}{=}}SEM \quad \overset{*}{\underset{2}{\rho}}=0{,}151 \qquad \underset{}{\overset{1}{=}}OX$$

The different substituents in $1a - 1x$ act in the expected way: with R = CH_3 ($1f$) as reference system electron donating substituents make reduction of 1 more difficult whereas electron attracting substituents hinder oxidation. For this reason $1a - 1f$, $1h - 1p$ and $1r$ are synthesized on the level "OX" whereas $1g$, $1q$ and $1s - 1x$ can be isolated only as form "RED".

Substituent effects on redox potentials can be rationalized by free energy relationships in many cases[8]. Although 1 can be related to aromatic compounds, a linear correlation with Hammett's σ-constants fails. As can be seen from the substituents $-N(CH_3)_2$ ($1k$) and $-OCH_3$ ($1l$) the electron attracting effect is the dominating one and not the resonance effect. Therefore Taft's σ^*-constants[9] describe the substituent effects more correctly. According to (1)

$$E - E_0 = \rho^* \, \Sigma \, \sigma^* \qquad \begin{aligned} &E = E_1 \text{ or } E_2 \\ &E_0 = E_1 \text{ or } E_2 \text{ for R} = CH_3 \; (1f) \end{aligned} \tag{1}$$

a linear correlation for E_1 as well as for E_2 is obtained (Fig. 1)[10]. The effects of the two N-substituents are additive as demonstrated by 1 with two different substituents, namely $1h$ (R = CH_3, $N(CH_3)_2$) : E_1 −0.60V, E_2 −0.23V; K_{SEM} $1.9 \cdot 10^6$ and $1p$ (R = $N(CH_3)_2$, $C_6H_3(NO_2)_2$) : E_1 0.20V, E_2 + 0.01 V; K_{SEM} $3.6 \cdot 10^3$.

By Eq. (1) unknown σ^* can be estimated as to be seen from Fig. 1 (σ^* of $SiMe_3$ uncertain because of irreversible potentials). The deviation of the charged substi-

Table 1. Potentials E_1 and E_2(V) of 1 in acetonitrile vs. Ag/AgCl/CH$_3$CN. K_{SEM} = Semiquinone formation constant; ( ) = irreversible

Comp.	R	E_1^a	E_2^a	ΔE	K_{SEM}
$1a$	$-\overset{\ominus}{C}(CO_2C_2H_5)_2$ [b]	-0.97	-0.59	0.38	$2.7 \cdot 10^6$
$1b$	$-\overset{\ominus}{C} \overset{CN}{\underset{CO_2C_2H_5}{}}$ [b]	-0.85	-0.63	0.22	$5.4 \cdot 10^3$
$1c$	$-\overset{\ominus}{C}(CN)_2$ [b]	-0.74	-0.60	0.14	$2.4 \cdot 10^2$
$1d$	$-CH(CH_3)_2$	-0.68	-0.27	0.41	$8.9 \cdot 10^6$
$1e$	$-CH_2CH_3$	-0.68	-0.28	0.40	$6.0 \cdot 10^6$
$1f$	$-CH_3$	-0.67	-0.26	0.41	$8.9 \cdot 10^6$
$1g$	$-Si(CH_3)_3$	$(-0.68)^c$	$(-0.45)^c$	(0.23)	$(7.9 \cdot 10^3)$
$1h$	$\underset{-N(CH_3)_2}{-CH_3}$	-0.60	-0.23	0.37	$1.9 \cdot 10^6$
$1i$	$-NH_2$	$-$	$(-0.27)^c$	$-$	$-$
$1j$	$-CH_2C_6H_5$	-0.60	-0.19	0.41	$8.9 \cdot 10^6$
$1k$	$-N(CH_3)_2$	-0.54	-0.17	0.37	$1.9 \cdot 10^6$
$1l$	$-OCH_3$	$-$	$(-0.19)^c$	$-$	$-$
$1m$	$-C_6H_5$	-0.37	-0.04	0.33	$3.9 \cdot 10^5$
$1n$	$CH(CN)CO_2C_2H_5$	-0.35	-0.62	0.33	$3.9 \cdot 10^5$
$1o$	$-CH(CO_2C_2H_5)_2$	-0.42	-0.07	0.35	$8.5 \cdot 10^5$
$1p$	$\underset{-C_6H_3(NO_2)_2}{-N(CH_3)_2}$	-0.20	$+0.01$	0.21	$3.6 \cdot 10^3$
$1q$	$-CON(CH_3)_2$	-0.07	$+0.09$	0.16	$5.2 \cdot 10^2$
$1r$	$-C_6H_3(NO_2)_2$	$+0.06$	$+0.19$	0.13	$1.6 \cdot 10^2$
$1s$	$-CO_2C_2H_5$	$+0.18$	$+0.35$	0.17	$7.6 \cdot 10^2$
$1t$	$-COCH_3$	$(+0.22)$	$(+0.25)$	0.03	$(3.2)^d$
$1u$	$-COC(CH_3)_3$	$+0.21$	$+0.27$	0.06	10^d
$1v$	$-COC_6H_5$	$+0.24$	$+0.32$	0.08	20^d
$1w$	$-CN$	$+0.47$	$+0.72$	0.25	$1.7 \cdot 10^4$
$1x$	$-COCF_3$	$(+0.68)^c$	$-$	$-$	$-$

a   Where the variation in DC-, AC- and CV-Measurements is less than 10 mV, the mean value is given. Where the variation is up to 20 mV, the mean of the two most similar values is given
b   Measured in DMF due to low solubility. Potentials are mostly less positive than in AN. To compensate, 0.01 V was added
c   Value from CV Measurement (20 V/s) only
d   Only the middle potential E_m was measured. K_{SEM} was estimated by standard methods[6], and used to obtain ΔE and hence E_1 and E_2

tuents in $1a$–$1c$ is not surprising. The effects of $-CO_2C_2H_5$ and $-CN$, however, are clearly at odds with expectation. Whereas their E_2-potentials are in line with σ^*, their E_1-potentials are much too low. This implies that in the electron transfer SEM/ RED these substituents cause an additional stabilisation, probably by resonance. But why don't acyl groups in $1t$-$1v$ act in the same way? This unusual behaviour of $-CO_2C_2H_5$ and $-CN$ is also observed in the corresponding derivatives of dipyridylethenes[11].

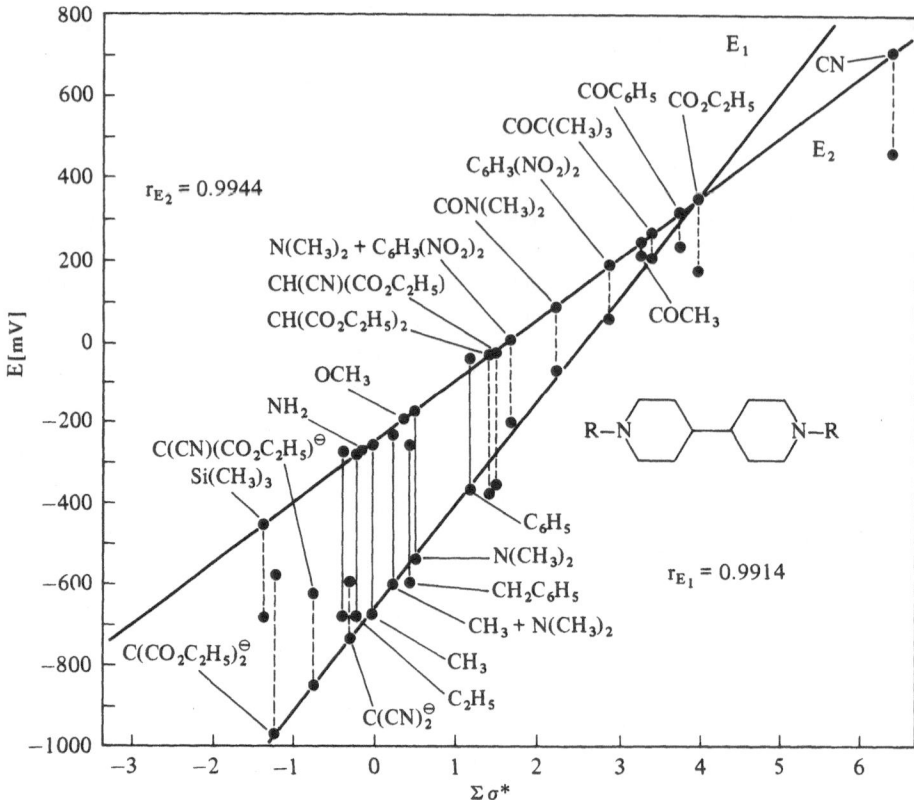

Fig. 1. Correlation of potentials E_1 and E_2 of *1* with the sum of the polar substituent constants σ^*. ●———●: σ^* from the literature; ●————●: σ^* this work[10]

As can be seen from Table 1 and Fig. 1, *the differences of E_1 and E_2 and there-fore K_{SEM} are reduced the more positive the potentials become.*

Correspondingly the linear plots for E_1 and E_2 in Fig. 1 converge since $\rho_1^* = 0.256$ is nearly twice as large as $\rho_2^* = 0.151$. Thus the redox step RED/SEM$^\oplus$ is much more sensitive to substituent effects than SEM$^\oplus$/OX$^{\oplus\oplus}$. This result is plausible since the polar influence of a substituent will be dominated by the larger amount of positive charge in the higher oxidation levels of *1*.

Based on R = CH$_3$ (*1f*) K_{SEM} is related to σ^* by Eq. (2)[10]

$$K_{SEM(1)} = 10^{6.95} \cdot 10^{-1.78 \Sigma \sigma^*} \tag{2}$$

For reasons already mentioned, (2) does not include the electrochemical properties of *1s* and *1w*.

From many of the discussed examples of *1* violet to green solutions of the radical ions could be obtained which are characterized by typical absorption bands[12]. Inter-estingly the longest wavelength absorption bands show a good linear correlation with

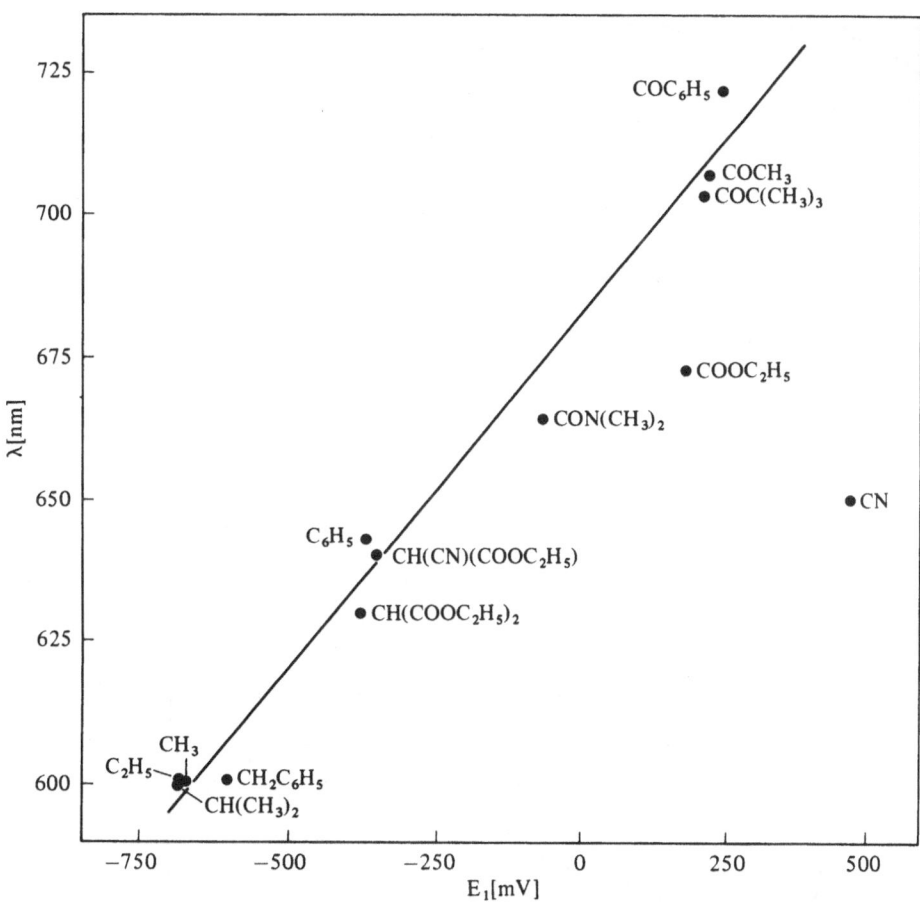

Fig. 2. Correlation of the longest wavelength absorption bands of 1_{SEM} in acetonitrile with E_1

E_1 or E_2 and thus also with σ^*, however again with exception of R = $CO_2C_2H_5$ and R = CN (Fig. 2).

This correlation underlines the indirect (polar) effect of most of the substituents on the π-system of 4,4'-bipyridyl. The largest difference observed (λ_{max} $1d$ = 603 nm vers. λ_{max} $1v$ = 722 nm) amounts to 32.2 kJ/mole.

2.2 Bisquaternary Salts of Isomeric Bipyridyls

One of the premises of the two step redox systems under discussion is a fully conjugated π-system on the reduced and oxidized level as represented by the general types A, B and C.

This presumption can be tested by the bipyridyl systems $1f$ and $2-4$.

7

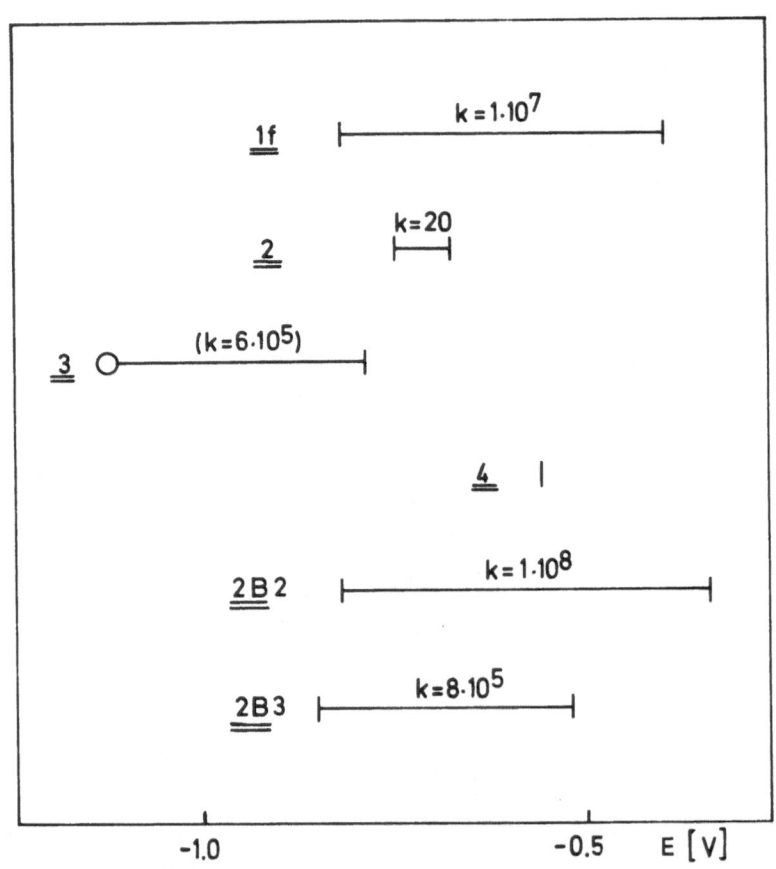

Ring connection	4,4′	2,2′	3,3′	3,4′
	$1f_{OX}$	2_{OX}	3_{OX}	4_{OX}

n	2	3
	$2B2_{OX}$	$2B3_{OX}$

Fig. 3. Potentials E_1, E_2(V) and K_{SEM} of $1f$, $2-4$, $2B2$ and $2B3$ in acetonitrile. vers. Ag/AgCl in sat. KCl[7]; ○ = irreversible electron transfer

As to be expected the ideal reversibility of the two step electron transfer in *1f* is lost if the pyridine rings are connected in 3,4' (*4*) and 3,3' (*3*) positions instead of 4,4' (*1f*). With *3* and *4* RED cannot be formulated without charges. The situation corresponds to that of the non-existing m-quinones. Not only does reduction in these cases occur at much more negative potentials but also irreversibly[7]. Thus the "instability" of the reduction products of *3* and *4* includes thermodynamic as well as kinetic properties. In contrast to *3* and *4* the 2,2'-isomer *2* undergoes completely reversible electron transfers, however at rather low potential with K_{SEM} of only ~20[7]. It was already presumed that the rather negative E_2 potential indicates the strongly twisted conformation of *2*[6e, 13a] which corresponds to that of 2,2'-bitolyl[13b]. If both rings are brought close to planarity[14] by a $(CH_2)_2$-bridge (*2B2*) potentials E_1 and E_2 move into the range of *1f* with an even larger K_{SEM} of $1 \cdot 10^8$ (herbicide "Diquat"®). The larger $(CH_2)_3$-bridge (*2B3*) twists the two pyridine rings distinctly causing K_{SEM} to drop to $8 \cdot 10^5$. It is not yet clear if the fixed syn-conformation of the two N-atoms in *2B2* and *2B3* reflects a special effect which is absent in the twisted anti-conformation of *2*. Calculated K_{SEM}'s for planar anti- and syn-conformations (cf. 6.1) are nearly equal[15].

2.3 Bisquaternary Salts of Isomeric Dipyridyl Ethenes

Lack of coplanarity in quaternary bipyridyls can be overcome not only by the discussed bridging but also by insertion of a vinylene group between the two pyridine rings.

The isomeric quaternary salts *5–10* again exemplify the validity of the general structure *A:* If the rings are connected to the vinylene group in 4,4'-, 2,2'- and 2,4'-positions (*5–7*) fully reversible redox reactions are observed with practically equal potentials and K_{SEM}'s[7]. (Fig. 4. The smaller K_{SEM}'s compared to *1f* are discussed in Sect. 5.)

Position of vinylene bridge	4,4'	2,2'	2,4'	2,3'	3,4'	3,3'
	*5*ox	*6*ox	*7*ox	*8*ox	*9*ox	*10*ox

If the vinylene bridge is attached in 3-position only E_2 is found to be a reversible electron transfer, which occurs at distinctly more negative potentials. This behaviour signalizes that in *8–10*, which cannot be derived from the general type *A*, the complete π-system is no longer involved. Of course this effect is most pronounced in *10* which contains two rings with "meta"-connections.

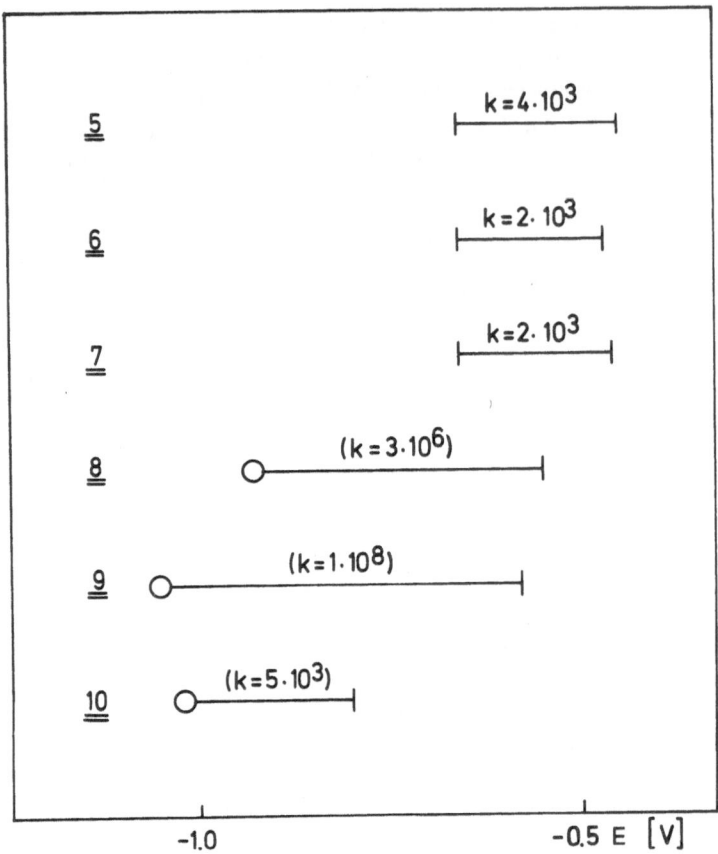

Fig. 4. Potentials E_1, E_2(V) and K_{SEM} of *5–10* in acetonitrile vers. Ag/AgCl in sat. KCl[7].
$\bigcirc$ = irreversible electron transfer

3 Bisquaternary Salts of Bis-aza Aromatics

3.1 Dimethyl-quaternary Salts of 3,8-Phenanthroline (*11*) and 2,7-Diazapyrene (*12*)

The classical Weitz-type redoxsystem *1f* can be transformed into *11* or *12*[16] by formal insertion of one or two vinylene groups. Although the π-system is considerably enlarged by this operation, E_1 and E_2 and therefore K_{SEM} are hardly changed (Fig. 5).

At first sight these results seem to be inconsistent with those of the corresponding hydrocarbons. These are more easily reduced with enlarged π-systems, namely biphenyl (E_2 −2.60V[17]), phenanthrene (E_2 −2.44V[18]) and pyrene (E_2 −2.08V[17]). Simple HMO-calculations demonstrate, however, that the high electron affinity of the N-atoms in *1f*, *11* and *12* changes the sequence of the MO's in such a way that symmetry as well as energy of the LUMO's are altered only to a small extent[19].

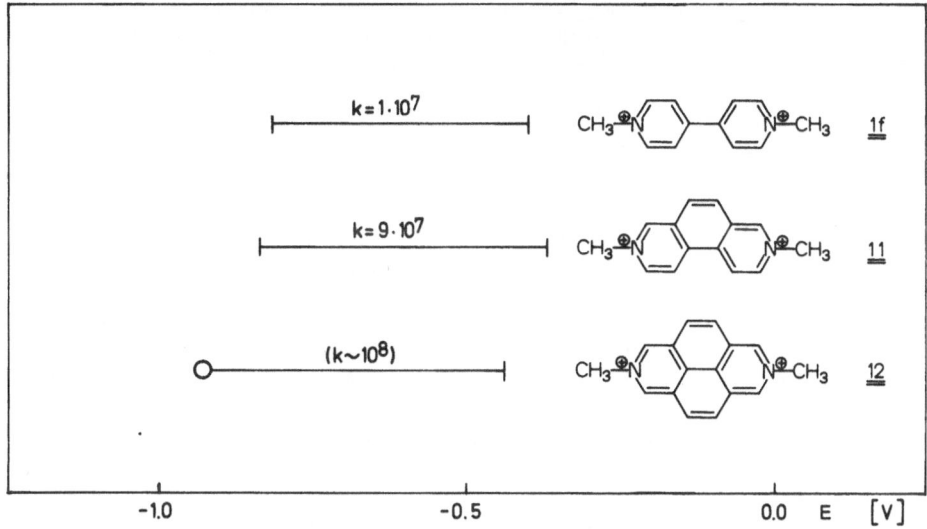

Fig. 5. Potentials E_1, E_2(V) and K_{SEM} of *1f, 11* and *12*[x] in acetonitrile versus Ag/AgCl in sat. KCl[16]. ([x] E_1 not reversible due to adsorption.)

3.2 Bisquaternary Salts of Isomeric Phenanthrolines

The bisquaternary salts *13*$_{OX}$ of the isomeric phenanthrolines were also investigated in order to test the generality of structure *A*. Again, a reversible two step reduction of these isomers is only to be expected if the two N-atoms are connected by an even number of C-atoms, so that *13*$_{RED}$ could be formulated in a "quinoid" structure. This is the case for *11, 13a–c, 13B2* and *13B3*, but not for *13d* and *13e*[16]. Isomers with $N–CH_3$ in position 1 had to be excluded because of rapid dimethylation even in organic solvents. In Table 2 results of rapid cyclic voltammetry[20] are accumulated, which correspond to those obtained earlier by DC-polarography[16].

Position of N-Atoms	3,8	4,7	2,9	2,7	3,7	2,8		n	2	3
	11$_{OX}$	*13a*$_{OX}$	*13b*$_{OX}$	*13c*$_{OX}$	*13d*$_{OX}$	*13e*$_{OX}$			*13B2*$_{OX}$	*13B3*$_{OX}$

As can be seen from Table 2 the situation is more complex than anticipated. Except for the already discussed *11*$_{OX}$ the electron transfer SEM/RED (E_2) is not or poorly reversible. Thus reliable K_{SEM}'s can only be derived from some of these potentials.

11

Table 2. Potentials E_1, $E_2(V)$ and K_{SEM} of isomeric quaternary phenanthrolines *11* and *13* in acetonitrile versus Ag/AgCl in acetonitrile[16]. Potentials in parentheses are irreversible

Nr. (OX)	Position of N-atom	$E_1{}^a$	$E_2{}^a$	K_{SEM}
11	3,8	−0.83	−0.35	$1.4 \cdot 10^8$
13a	4,7	(−0.74)	−0.35	$(4 \cdot 10^6)$
13b	2,9	$(−0.79)^b$	$−0.42^b$	$(1 \cdot 10^6)$
13c	2,7	$(−0.77)^b$	$−0.35^b$	$(9 \cdot 10^6)$
13d	3,7	(−1.08)	−0.66	$(2 \cdot 10^8)$
13e	2,8	(−1.12)	−0.65	$(3 \cdot 10^8)$
*13*B2	1,10	$(−0.74)^c$	−0.30	$(1.4 \cdot 10^8)$
*13*B3	1,10	$(−0.71)^c$	−0.27	$(9 \cdot 10^7)$

^{a} CV with 20 V/s ^{b} Slow heterogenous charge transfer ^{c} Strong adsorption

Nevertheless, positions and differences of E_1 and E_2 allow some qualitative interpretations. The vinylene bridge in 5,6-position seems to be "inert" not only in *11*, which resembles the 4,4′-bipyridyl system *1f* (c.f. 3.1) but also in *13*B2 and *13*B3 which behave similar to the 2,2′-bipyridyl systems *2*B2 and *2*B3. In *13*B3 the vinylene group prevents the two pyridyl units from distortion by the C_3-bridge between the nitrogen atoms compared to *2*B3. Thus its K_{SEM} stays closer to that of *13*B2.

In $13a,b,c_{OX}$ the first electron (E_2) is accepted rather easily, so that these isomers can be connected to the corresponding dipyridylethylenes *6*, *5* and *7*. This interpretation is supported by the fact that the only two final isomers $13d_{OX}$ and $13e_{OX}$ are reduced much more difficultly.

Table 3. Potentials E_1, $E_2(V)$ and K_{SEM} of isomeric naphthyridine bismethylquaternary salts 14_{OX} in acetonitrile versus Ag/AgCl in sat. KCl[21]. Potentials in parentheses are irreversible

$$CH_3 - \overset{\oplus}{N} \overset{}{=} \overset{\oplus}{N} - CH_3$$

Nr. (OX)	Position of N-atoms	$E_1{}^a$	$E_2{}^a$	K_{SEM}
14a	1,5	−0.47	+0.21	$8.9 \cdot 10^{11}$
14b	1,7	−0.63	+0.20	$2.5 \cdot 10^{14}$
14c	2,6	−0.67	+0.12	$1.4 \cdot 10^{13}$
14d	1,6	$(−1.5)^x$	(+0.13)	$(4 \cdot 10^{27})$
14e	2,7	$(−1.2)^x$	$(−0.14)^x$	−
14f	1,8	−	$(+0.32)^x$	−
$14g^b$	1,8	−	$(+0.22)^x$	−
$14h^c$	1,8	$(−1.4)^x$	$(+0.20)^x$	−

^{a} Scan rate 200 mV/s ^{b} $−(CH_2)_2−$ instead of $−CH_3$ ^{c} $−(CH_2)_3−$ instead of $−CH_3$. ^{x}: $E_{1/2}$ measured at 85% ip_c

3.3 Bisquaternary Salts of Isomeric Naphthyridines

Further iso-π-electronic systems of closely related geometry can be derived from the isomeric naphthyridines. Their bismethyl quaternary salts 14_{OX}[21] exhibit the expected electrochemical properties (Table 3). Only the isomers $14a-c_{OX}$ accept the first *and* the second electron reversibly. As expressed by K_{SEM} $10^{12} - 10^{14}$ the thermodynamic stability of the corresponding radical cations is unusually high. Electrochemical data about the other isomers, $14d-h_{OX}$, can be obtained only by rapid cyclic voltammetry (20 V/s) and even then reversibility is lacking. Due to the rigidity of the system potentials E_2 of $14g$ and $14h$ differ very little.

4 Variation of Heteroatoms

Naturally the redox properties of the general structures $A-C$ should strongly be influenced by different heteroatoms X. Since this report deals mainly with *Weitz* type systems, X is incorporated into a heterocyclic ring.

4.1 Six-membered Heterocycles

The basic structure of the 4,4'-bipyridinium-salts 1_{OX} is also included in the iso-π-electronic bipyrylium and bithiopyrylium salts 15_{OX} and 16_{OX}[22]. The more positive potentials are already expressed in the synthetic route which leads initially to 15_{RED} and 16_{RED}[22].

Obviously the higher electron affinity of O and S compared to N in $1f$ provides the systems 15 and 16 with electrochemical properties similar to 1 with strongly electron attracting substituents: The more positive potentials are connected to a smaller ΔE; i.e. K_{SEM} is reduced by two powers of ten.

Table 4. Potentials E_1, E_2(V) and K_{SEM} of $1f$, 15 and 16 in acetonitrile versus Ag/AgCl in sat. KCl[22]

Nr.	X	E_1	E_2	K_{SEM}
$1f_{OX}$	$N-CH_3$	−0.80	−0.39	$8.9 \cdot 10^6$
15_{OX}	O	+0.21	+0.50	$8.2 \cdot 10^4$
16_{OX}	S	+0.20	+0.46	$2.5 \cdot 10^4$

Table 5. Longest wavelength absorption maxima (nm) and coupling constants a in α- and β-position of radical cations $1f_{SEM}$[24], 15_{SEM} and 16_{SEM}[23a] in acetonitrile

	$1f_{SEM}$	15_{SEM}	16_{SEM}
X	N–CH$_3$	O	S
λ_{max}	735	596	727
a_α	1.33	2.97	2.36
a_β	1.57	0.80	0.58

The corresponding radical cations are kinetically highly stable so that UV-[22] and ESR-[23] spectra can easily be recorded. As to be seen from Table 5 the absorption maxima of the radical ions are similar with X = N–CH$_3$ ($1f_{SEM}$) and S (16_{SEM}) whereas O in 15_{SEM} causes a strong hypsochromic shift. In ground state properties *15* and *16* are more closely related as, for instance, with regard to the redox potentials and the coupling constants a of the radical ions. Here the already known sequence $a_\beta > a_\alpha$ for $1f_{SEM}$[24] is reversed parallel with the increasing electronegativity of the heteroatoms. The assignment was made by comparison with the corresponding radical cations methylated in the α-position[23a]. For 16_{SEM} identical values have been measured and assigned by McConnel's equation[23b].

4.2 Five-membered Heterocycles

Five-membered heterocycles with two heteroatoms connected in 2-position also form two step redox systems of the *Weitz* type which may be described as heterotetrasubstituted ethylenes on the level of "RED". The great significance of tetrathia (selena) fulvalenes *17* as a basis for "organic metals" has already been discussed[1, 25]. In this rapidly developing field the crystal structure of the specific donor-acceptor pair dominates the solution bound redox potentials which are discussed here.

Table 6 summarizes fulvalenes of type *17* as far as their redox potentials have been measured.

The long-known dibenzo-tetrathiafulvalene 18_{RED}[35] can also be oxidized reversibly in two separate steps with E$_1$ = +0.72 V, E$_2$ = +1.06 V (vers. Ag/AgCl in acetonitrile) and K$_{SEM}$ = 5.6 · 10^5 [28]. The influence of different heteroatoms can be read from the systems $19-21$[36], which are better isolated in the oxidized form.

X	O	S	N–CH$_3$
	19_{OX}	20_{OX}	21_{OX}

Table 6.

R^1, R^2, X, Y, R^3, R^4 — structure **17**

Potentials E_1, E_2(V) and K_{SEM} of fulvalenes *17* versus saturated calomel electrode

Nr.	R^1	R^2	R^3	R^4	E_1^c	E_2^c	K_{SEM}	Ref.
17A								
$X=Y=S$								
17Aa	H	H	H	H	0.33	0.70	$1.8 \cdot 10^6$	26, 27)
					0.45^a	0.82^a	$1.8 \cdot 10^6$	28)
					0.30	0.66	$1.3 \cdot 10^6$	29)
					0.34	0.71	$1.8 \cdot 10^6$	30)
17Ab	CH_3	CH_3	CH_3	CH_3	0.27			31)
17Ac	$-(CH_2)_3-$		$-(CH_2)_3-$		0.26^b	0.67^b	$8.9 \cdot 10^6$	32)
					0.33	0.66	$3.9 \cdot 10^5$	33)
17Ad	$-(CH_2)_4-$		$-(CH_2)_4-$		0.25^b	0.67^b	$1.3 \cdot 10^7$	32)
					0.40^a	0.80^a	$6.3 \cdot 10^6$	28)
17Ae	$-(CH_2)_5-$		$-(CH_2)_5-$		0.21^b	0.64^b	$1.9 \cdot 10^7$	32)
17Af	$-CH(CH_3)CH_2CH_2-$		$-CH(CH_3)CH_2CH_2-$		0.30	0.64	$5.7 \cdot 10^5$	33)
17Ag	$-CH_2CH(CH_3)CH_2-$		$-CH_2CH(CH_3)CH_2-$		0.30	0.63	$4.0 \cdot 10^5$	33)
17Ah	$-SCH_2CH_2-$		$-CH_2CH_2S-$		0.43	0.69	$2.6 \cdot 10^4$	33)
17Ah'	$-SCH_2CH_2-$		$-SCH_2CH_2-$		0.43	0.69	$2.6 \cdot 10^4$	33)
17Ai	C_6H_5	H	C_6H_5	H	0.39^b	0.78^b	$4.1 \cdot 10^6$	32)
17Aj	C_6H_5	CH_3	C_6H_5	CH_3	0.34^b	0.73^b	$4.1 \cdot 10^6$	32)
17Ak	C_6H_5	C_6H_5	C_6H_5	C_6H_5	0.41^b	0.77^b	$1.3 \cdot 10^6$	32)
17Al	$4-CH_3O-C_6H_4$	H	$4-CH_3O-C_6H_4$	H	0.33^b	0.73^b	$6.0 \cdot 10^6$	32)
17Am	$4-CH_3-C_6H_4$	H	$4-CH_3-C_6H_4$	H	0.35^b	0.76^b	$8.9 \cdot 10^6$	32)
17An	$4-Br-C_6H_4$	H	$4-Br-C_6H_4$	H	0.43^b	0.78^b	$8.5 \cdot 10^5$	32)
17Ao	$4-Cl-C_6H_4$	H	$4-Cl-C_6H_4$	H	0.43^b	0.79^b	$1.3 \cdot 10^6$	32)
17Ap	Br	H	Br	H	0.59			31)
17Aq	COOH	H	H	H	0.47	0.83	$1.3 \cdot 10^6$	30)
17Ar	COOEt	H	H	H	0.47	0.83	$1.3 \cdot 10^6$	30)
17As	COOEt	H	COOEt	H	0.60	0.84	$5.8 \cdot 10^5$	30)
17At	$COOCH_3$	$COOCH_3$	$COOCH_3$	$COOCH_3$	0.80	1.08	$5.6 \cdot 10^4$	30)
					0.80			31)
17Au	CF_3	CF_3	CF_3	CF_3	1.08			31)
17Av	CN	CN	CN	CN	1.12	1.22	49	34)
					1.12			31)
17B								
$X=S, Y=Se$								
17Ba	H	H	H	H	0.40	0.72	$2.7 \cdot 10^5$	27)
17Bd	$-(CH_2)_4-$		$-(CH_2)_4-$		0.28	0.63	$8.5 \cdot 10^5$	29)
17Bg	$-CH_2CH(CH_3)CH_2-$		$-CH_2CH(CH_3)CH_2-$		0.39	0.70	$1.8 \cdot 10^5$	33)
17C								
$X=Y=Se$								
17Ca	H	H	H	H	0.48	0.76	$5.5 \cdot 10^4$	27)
17Cb	CH_3	CH_3	CH_3	CH_3	0.44	0.72	$5.5 \cdot 10^4$	33)
17Cg	$-CH_2CH(CH_3)CH_2-$		$-CH_2CH(CH_3)CH_2-$		0.48	0.75	$5.9 \cdot 10^4$	33)

^{a} Vs. Ag/AgCl/CH_3CN ^{b} In MeOH/Benzene (4:1)/0.1 M LiCl ^{c} In CH_3CN/0.1 M TEAP

Table 7. Potentials E_1, E_2(V) and K_{SEM} for systems *19–21* in acetonitrile versus Ag/AgCl in sat. KCl[36]

Nr.	E_1	E_2	K_{SEM}
19$_{OX}$	−0.29	+0.18	$9.1 \cdot 10^7$
20$_{OX}$	−0.17	−0.02	$3.5 \cdot 10^2$
20B$_{OX}$	−0.23	+0.26	$2 \cdot 10^8$
21$_{OX}$	−0.91	−0.84	~16
21B$_{OX}$	−1.32	--0.47	$5 \cdot 10^{14}$

All compounds can be reversibly reduced in two steps (Table 7). Thermodynamic stability of *19*$_{SEM}$ with $K_{SEM} = 1 \cdot 10^8$ is rather high. The unexpected decrease of K_{SEM} on substitution of X = O (*19*) by X = S (*20*) and especially X = N–CH$_3$ (*21*) has to be attributed to disturbed coplanarity: By restoring planarity the undisturbed π-system in the bridged compounds *20B*$_{OX}$ and *21B*$_{OX}$ is more easily reduced (potentials more positive) and K_{SEM} is strongly increased. The large K_{SEM} of *20B* seems not to depend on the fixed Z-configuration since calculations for the planar E-configuration of *20* produce the same K_{SEM} (cf. 2.2. and 5.1.). As to be expected *21B*$_{OX}$, being a derivative of the strongly basic benzimidazole, is the system most difficult to reduce. The influence of the heteroatoms in *19–21* reflects that in the six-membered heterocycles *1f, 15* and *16*.

5 Vinylogous Redox Systems

The effect of an increasing number of vinylene groups in the general type *A* should be of special interest. The separation of the end groups X which carry an appreciable amount of positive charge on the level SEM and OX increases as the π-system is enlarged.

22_{OX} to 30_{OX} have been synthesized as vinylogous series of *Weitz* systems some members of which have already been discussed under a different point of view. For comparison 31_{RED} is mentioned as an example of the "inverse Weitz type"[1] and 32_{OX} to represent the "open chain type"[1]. In all cases an all-E-configuration has to be assumed, although rapid E/Z equilibration via their redox equilibria cannot be excluded[37].

22_{OX} n = 0–3

23_{OX} n = 0–3

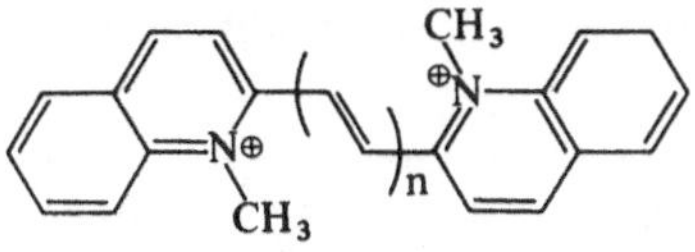

24_{OX} n = 0—3

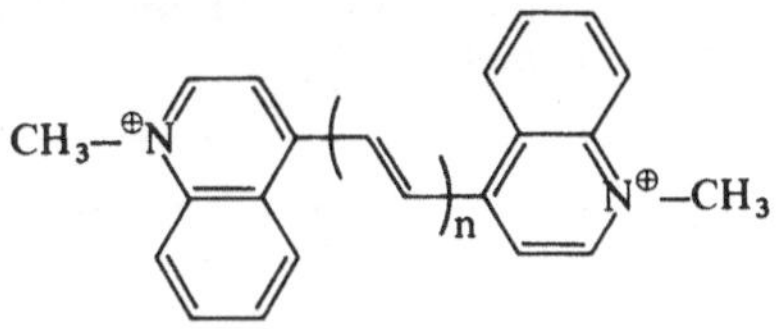

25_{OX} n = 0—3

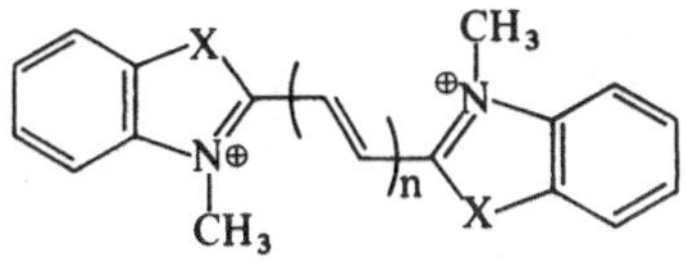

X	O	S	N—CH$_3$
	26_{OX}	27_{OX}	28_{OX}
	n = 0—2	n = 0—3	n = 0—3

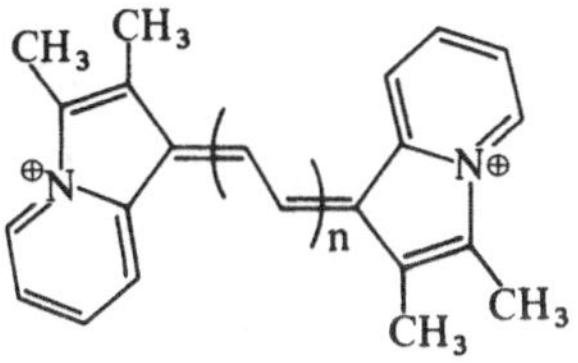

29_{OX} n = 0—4

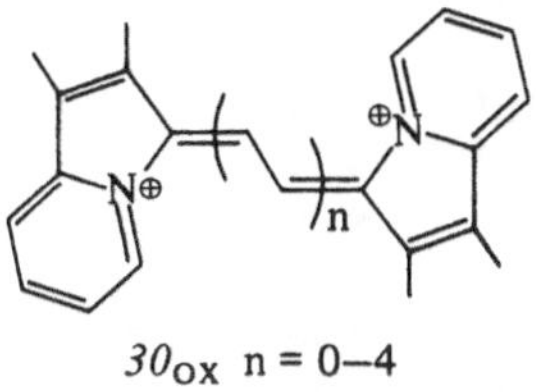

30_{OX} n = 0—4

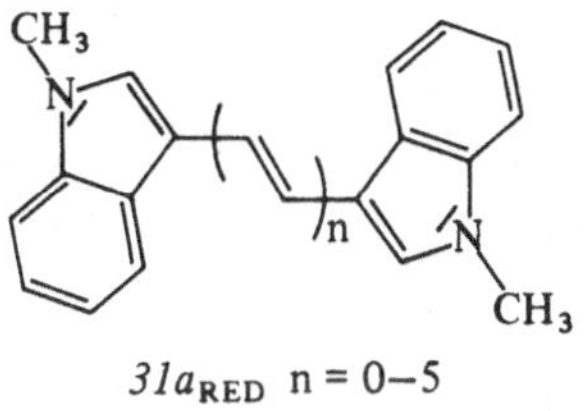

$31a_{RED}$ n = 0—5

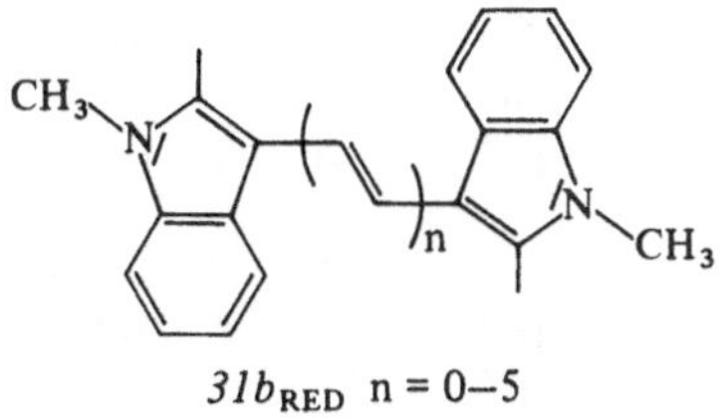

$31b_{RED}$ n = 0—5

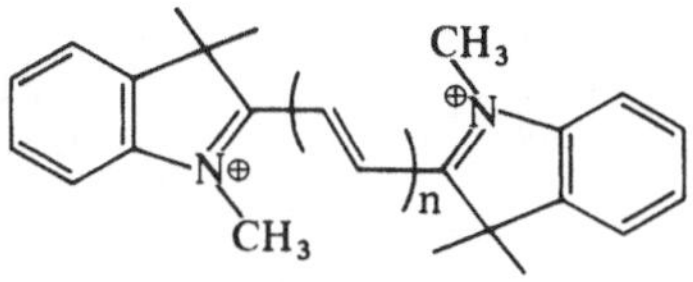

32_{OX} n = 1—5

5.1 Redox Properties

All the vinylogous systems *22—32* exhibit the same pattern of redox properties. The position of the potentials E_1 and E_2, or better $E_m = (E_1 + E_2)/2$ of a certain vinyl-ogue is mainly determined by the heterocyclic end groups (cf. Table 8).

17

Table 8. Potentials E_1, E_2 or E_m(V) and K_{SEM} of vinylogous systems *22–32* in acetonitrile and Coulomb repulsion integral J_{mm}^{SEM}

Nr.	n	E_1	E_m	E_2	K_{SEM}	J_{mm}^{SEM}	Ref.
*22*OX[a]	0(=*2*)	−0.63		−0.55	28		38)
	1(=*6*)	−0.52		−0.33	$1.5 \cdot 10^3$	4.071	
	2		−0.42		2.6	3.679	
	3		−0.41		$4.8 \cdot 10^{-2}$	3.369	
*23*OX[a]	0(=*1f*)	−0.68		−0.27	$1.3 \cdot 10^7$	4.407	38)
	1(=*5*)	−0.54		−0.34	$2.5 \cdot 10^3$	3.929	
	2		−0.43		1.3	3.551	
	3		−0.42		$4.8 \cdot 10^{-2}$	3.252	
*24*OX[a]	0	−0.25		−0.11	$1.9 \cdot 10^2$	4.164	38)
	1	−0.16		−0.05	90	3.808	
	2		−0.13		$4 \cdot 10^{-1}$	3.498	
	3		−0.13		$4.8 \cdot 10^{-2}$	3.235	
*25*OX[a]	0	−0.35		−0.28	21	4.048	38)
	1	−0.18		−0.10	30	3.694	
	2		−0.16		$3.2 \cdot 10^{-1}$	3.387	
	3		−0.18		$4.8 \cdot 10^{-2}$	3.134	
*26*OX[b]	0(=*19*)	−0.29		+0.18	$9.1 \cdot 10^7$	5.21	39)
	1	−0.22		+0.09	$2.0 \cdot 10^5$	4.57	
	2	−0.18		−0.08	50	4.06	
*27*OX[b]	0(=*20*)	−0.17		−0.02	$3.5 \cdot 10^2$	4.80	39)
	1	−0.19		+0.08	$4.0 \cdot 10^4$	4.30	
	2	−0.14		−0.07	~14	3.87	
	3	−0.17		−0.12	~ 7	3.53	
*28*OX[b]	0(=*21*)	−0.91		−0.84	~16	5.10	39)
	1	−0.76		−0.61	$3.5 \cdot 10^2$	4.50	
	2	−0.69		−0.62	$1.5 \cdot 10^2$	4.01	
	3		−0.62		−		
*29*OX[b]	0	+0.43		+0.69	$1.1 \cdot 10^4$		40)
	1	+0.05		+0.20	$3.5 \cdot 10^2$		
	2		+0.08		11(32[c])		
	3		+0.06		(5[c])		
	4		+0.03		(0.3[c])		
*30*OX[b]	0	+0.43		+0.67	$1.1 \cdot 10^4$		40)
	1	+0.07		+0.22	$3.2 \cdot 10^2$		
	2		+0.09		21		
	3		+0.06				
	4		+0.04				

Table 8. (continued)

Nr.	n	E_1	E_m	E_2	K_{SEM}	J_{mm}^{SEM}	Ref.
$31a_{RED}{}^a$	0	+0.66		+1.03	$4.1 \cdot 10^6$		37)
	1	+0.52		+0.75	$7.9 \cdot 10^3$		
	2	+0.51		+0.56	~ 10		
	3		+0.47		$\sim 7 \cdot 10^{-1}$		
	4		+0.43				
	5		+0.40				
$32_{OX}{}^b$	1	−0.06		+0.28	$5.8 \cdot 10^5$		37)
	2	−0.06		+0.09	$3.2 \cdot 10^2$		
	3		−0.03		3		
	4		−0.05		$(2 \cdot 10^{-2}\ ^c)$		
	5		−0.07		$(3 \cdot 10^{-3}\ ^c)$		

^{a} Potentials vers. Ag/AgCl in acetonitrile ^{b} Potentials vers. Ag/AgCl in sat. KCl
^{c} By UV/VIS-spectroscopy

Since aromaticity is lost on reduction, quinoline derivatives 24_{OX} and 25_{OX} are reduced more easily than the pyridine systems 22_{OX} and 23_{OX} (compare also 5_{OX} and 6_{OX}). Only in 31_{RED} the situation is reversed. On oxidation it *loses* aromaticity and therefore is most difficult to oxidize (potentials rather positive). 32 in which aromaticity is neither lost nor gained during the redox process serves as a "neutral" example.

With increasing numbers of vinylene groups partly the positions of E_1 and partly those of E_2 are more influenced. The reason for this phenomenon is not clear. Different solvation energies may be involved.

The most important result in Table 8, however, concerns K_{SEM}: *The longer the vinylene chain the smaller K_{SEM} becomes.* The highest K_{SEM} in each series is always connected with the first member, provided that it is planar. These vinylogous series confirm that the much too low K_{SEM}'s of (n = 0) systems, 22 (=2), 24, 27 (=20) and 28 (=21) are due to lack of planarity. Therefore they have to be substituted by the planar derivatives, e.g. $2B2$, $20B$ and $21B$.

Qualitatively the decrease of K_{SEM} with increasing chain length can be interpreted by separation of the positive charges. SEM therefore will more easily gain or loose an electron, i.e. ΔE and correspondingly K_{SEM} will become smaller.

Quantitatively the Coulomb repulsion integral J_{mm}^{SEM} of SCF-calculations for SEM correlates linearly with lg K_{SEM} [41]. Equation (3)

$$J_{mm}^{SEM} \approx \Delta H \text{ for } \Delta H = 2H_{SEM}^\circ - (H_{RED}^\circ - H_{OX}^\circ) \tag{3}$$

holds true inspite of drastic variations in the end groups (Fig. 6). Even the redox properties of some polyenes without any heteroatoms [15, 42] which correspond to type A or B are governed by this correlation. Therefore thermodynamic stabilities

19

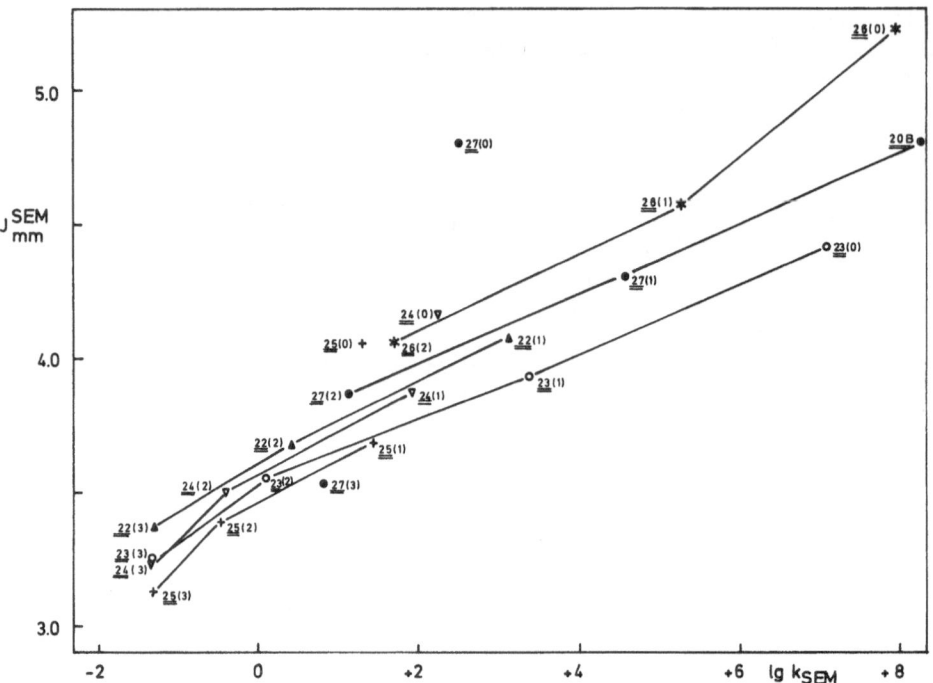

Fig. 6. Correlation between lg K_{SEM} of different vinylogous redox systems and the Coulomb repulsion integral J_{mm}^{SEM}

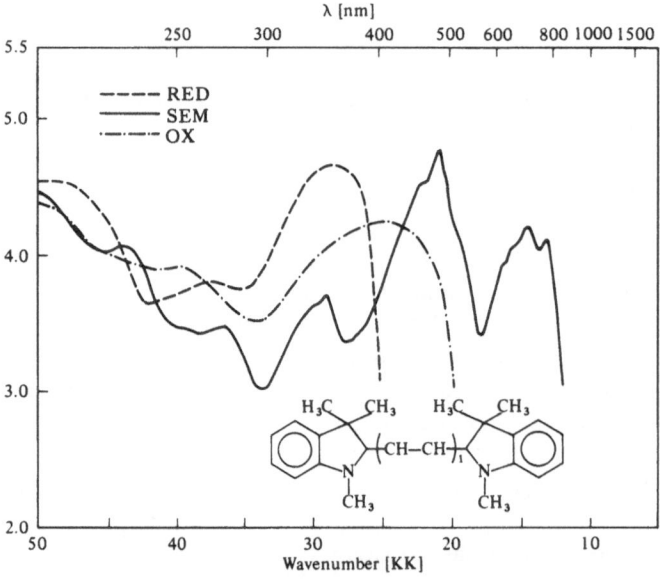

Fig. 7. UV/VIS-spectra of 32_{RED}, 32_{SEM} and 32_{OX} (n = 1) in acetonitrile[44]

of different SEM's can be predicted quite reasonably from Eq. (3). Of the above mentioned planar substitutes for n = 0 *20B* fits well the linear correlation in Fig. 6, in contrast to the much too small K_{SEM} of *27*(0) (also compare *24*(0)). When the electrochemical data are difficult to obtain, deviations are more serious (e.g. *27*, n = 3). K_{SEM}'s smaller than ~20 have to be determined by UV/VIS spectrometry which can only be applied if all three partners of a given redox system are kinetically reasonably stable. This is the case for *31*[37] and *32*[43, 44].

5.2 A Colour Rule for Vinylogous Radical Cations ("Violenes"[2])

From *Wurster's* salts[45] and the *Weitz* system *1f* it is well known that the intermediate oxidation level of the radical cations ("violenes") is characterized by an especially long wavelength absorption. This general behaviour may be demonstrated with *32*, n = 1[44]. In Fig. 7 the similar absorption curves of RED and OX can be seen from which the oxidation level SEM differs totally by its strong bathochromic shift combined with two strong absorption maxima. Calculations on *32*$_{SEM}$ (n = 1) as an open shell system[44], reveal that these two bands are connected to the two highest energy transitions (HOMO ⟶ SOMO and SOMO ⟶ LUMO) which are mixed by configuration interaction. The longest wavelength band represents to ~67% the transition HOMO ⟶ SOMO (Fig. 8).

Already before systematical investigations were started, it was anticipated that the known "vinylene shift" of 90–120 nm of the "cyanine type" may also be found in the "violene type"[44]. This situation has subsequently been expressed in a general theory of polymethines[2a, b].

In both cases the number of π-electrons differs from that of the sp$_2$-atoms by one. Accordingly a linear correlation is expected between the absorption maxima of cyanines and violenes, and the number of vinylene groups or π-electrons.

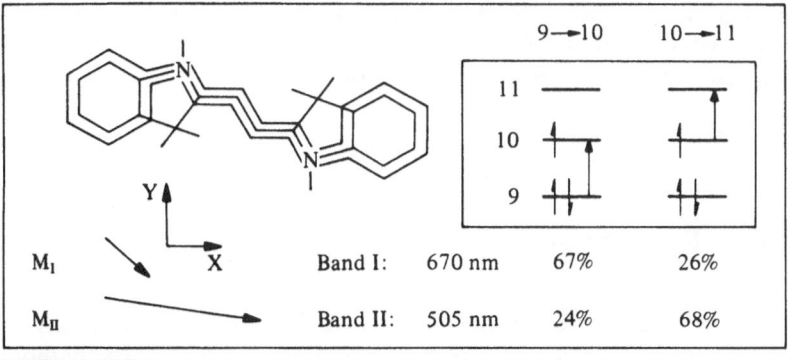

Fig. 8. Relative amounts of exitation in energy levels 9 ⟶ 10 and 10 ⟶ 11 in the long wave length bands of *32*$_{SEM}$ (n = 1) together with the transition moments M_I and M_{II}[44]

Cyanine Type	Violene Type
2n + 4 π-electrons on 2n + 3 sp²-atoms	2n + 3 π-electrons on 2n + 2 sp²-atoms

$$\text{Cyanine Type: } 2n + 4 \; \pi\text{-electrons on } 2n + 3 \; sp^2\text{-atoms}$$

$$\text{Violene Type: } 2n + 3 \; \pi\text{-electrons on } 2n + 2 \; sp^2\text{-atoms}$$

Quantitative MO-LCAO calculations[2a] as well as treatments as a one dimensional electron gas[46] have been advanced. Special parameters, however, have to be introduced to account for different end groups and branching of the π-system. Empirically a linear correlation between λ_{max}^{SEM} and n is verified in all cases so far investigated. That is, *violenes behave like cyanines*. The vinylene shift amounts to 100—150 nm in contrast to that of the corresponding forms OX and RED with 20—40 mm[44].

5.3 Correlation of λ_{max} of Violenes with the Number of π-Electrons

Equation (4), which corresponds to the well known correlation for cyanines[47], describes λ_{max} of vinylogous violenes extremely well.

$$\lambda_{max}^{SEM} = k \cdot N + b \qquad (4)$$

k, b = constants   N = N = number of π-electrons within the shortest distance of the end groups

The specific effect of the different end groups is expressed by variations of k (branching, heteroatom) and b (additional π-electrons)[38]. As can be read from Table 9, the correlation coefficient r is close to 1.0 in most cases. In some cases for n = 0 there are deviations which obviously arise from steric hindrance[48]. Thus these data are omitted from the correlation. With cyanines one finds k ~65 which some of the violenes exhibit too. Figure 9 again demonstrates the discussed correlation.

6 Aza-vinylogous Redox Systems

In the field of cyanine dyes aza-substitution has been studied extensively for nearly 100 years and even a theoretical treatment of the observed hypsochromic and bathochromic shifts is possible[46, 49]. In case of the violenes with their even number of methine groups between the two ends, symmetrical bis- or tetraaza-substitution is most suitable both for practical and theoretical (electrochemical) reasons.

Table 9. Correlation of λ_{max}^{SEM} of redox systems carrying different end groups, with the number of π-electrons $N = 2n + 3$ for vinylogous n (n in parantheses excluded)

System No.	End-group	n	k	b	r
22		0–3	66.00	+133.00	0.9998
23		(0) 1–3	72.50	−15.16	0.99996
24		1–3	53.50	+505.50	0.9994
25		(0) 1–3	82.25	+28.08	0.9997
27		0–3	67.75	+210.25	0.9997
32		1–5	64.15	+310.15	0.996
		(1) 2–5	67.85	+262.05	0.997
29		1–4	57.10	+494.30	0.9997
30		1–4	73.00	+25.00	0.9988
		(1) 2–4	77.00	−45.00	0.9995
31a		0–2	52.50	+359.16	0.994
31b		1–3	57.25	+312.58	0.995

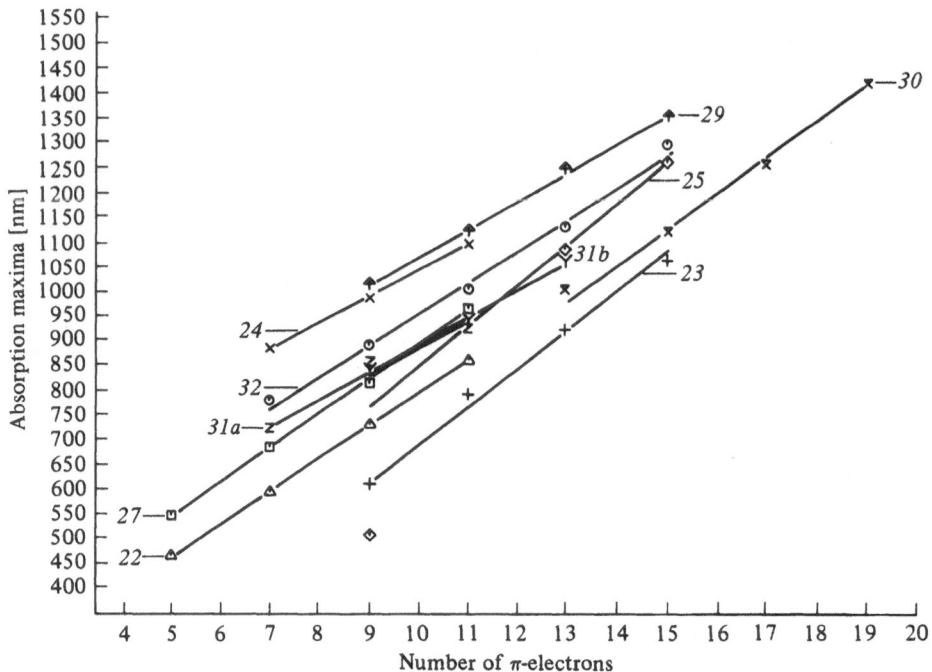

Fig. 9. Correlation of λ_{max}^{SEM} of redox systems carrying different end groups with the number of π-electrons N = 2n + 3 for vinylogous n

6.1 Redox Systems with a Diaza-dimethine Chain

Coupling of two heterocycles by a N—N-bridge is simple to achieve and produces especially significant examples of two step redox systems with excellent reversibility. With the following compounds *33—39* an increase in K_{SEM} of 10^4-10^5 compared to the dimethine derivatives is observed, so that most K_{SEM}'s are $\sim 10^{10}$. Additionally all potentials become more positive by as much as ~ 1 V. Besides with hardly any exception, all three oxidation levels of these systems can be isolated[39, 50, 51].

(RED)	*33*	*34*	*35*	*36*	*37*
X	NCH_3	O	S	Se	$C(CH_3)_2$
K_{SEM}	$3.0 \cdot 10^{10}$	$1.0 \cdot 10^{10}$	$6.5 \cdot 10^8$	$3.0 \cdot 10^8$	$1.5 \cdot 10^{10}$

$\underline{\underline{33}}-\underline{\underline{37}}$ RED $\underline{\underline{38}}$RED $\underline{\underline{39}}$RED

The isomeric indolizines *38* and *39* seem to contradict these rules: Compared to their dimethine derivatives *30* (n = 1) and *29* (n = 1) potentials are more positive by ~0.1 V only and their K_{SEM}'s are quite similar. In these cases, however, the reduced form contains an azo group which changes to the less energetic azino group[40] on oxidation, simultaneously transforming the indolizine into a pyridinium ring.

This aza-effect is preserved if in a C_4-bridge the two central C-atoms are aza-substituted as in *40*$_{RED}$[39] (cf. *28*, n = 2). Even four nitrogens in a row as in *41*$_{RED}$ are tolerated[39] and act in the same way.

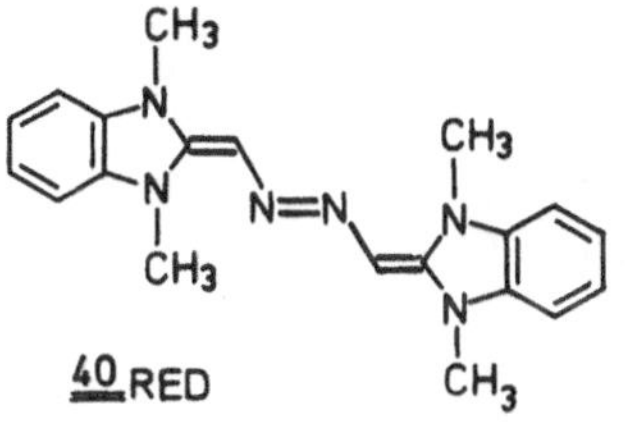

40$_{RED}$

$E_1 = -0.60 \; E_2 = -0.24 \; K_{SEM} = 1.0 \cdot 10^6$

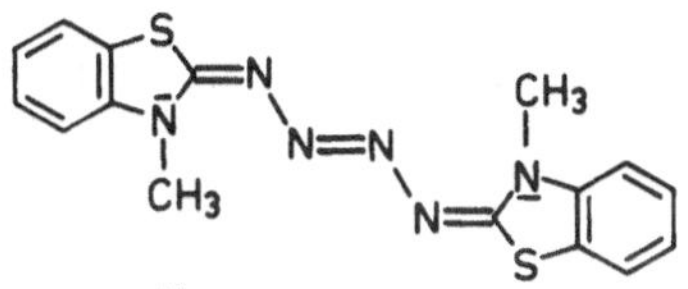

41$_{RED}$

$E_1 = +0.93 \; E_2 = +1.21 \; K_{SEM} = 5.5 \cdot 10^4$

7 Redox Systems with Cyclic π-Bridges

Instead of vinylene bridges and their aza-analogues the heterocyclic end groups of a *Weitz* system can be linked by iso- or heterocyclic π-ring systems. The remarkable variability of these cyclic bridges is demonstrated by the following examples.

7.1 Isocyclic Three-membered Rings as Bridges

The cyclopropene *42*$_{OX}$ can be regarded as a derivative of *5*$_{OX}$ in which the ethylene bridge has been forced into a three-membered ring. Thereby the two pyridine rings are fixed in a cis-configuration. Despite these special arrangements K_{SEM} of *42* is close to that of *5*, however reduction occurs more easily (~0.3 V[52]). A possible explanation is given below.

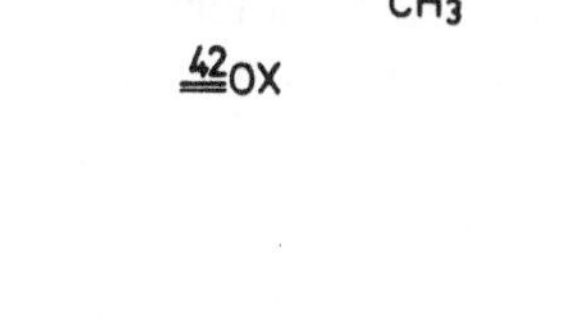

42$_{OX}$

E_1	E_2	K_{SEM}[a]
-0.32V	-0.11[b]	~$5 \cdot 10^3$

[a] In acetonitrile vs. Ag/AgCl/CH_3CN; CV 20V/s

[b] Irreversible electron transfer

7.2 Isocyclic Four-membered Rings as Bridges

7.2.1 Cyclobutene Derivatives as Bridges

A situation similar to 42_{OX} is depicted in 43_{OX}[11] in which the bond angles of the vinylene bridge are less distorted than in 42_{OX}. Thus hybridisation in 43 is more

E_1	E_2	K_{SEM}[a]
−0.40V	−0.24V	$6.3 \cdot 10^2$

[a] In acetonitrile vs. Ag/AgCl/CH$_3$CN

similar to that in the dipyridylethene system 5 than that of 42. At least the shift to more negative potentials in the sequence $42 < 43 < 5$ parallels the alteration of hybridisation[53]. The extra positive charges in 42 and 43 seem to be less important.

7.2.2 Four Step Electron Transfer Between a [4] Radialene and a Cyclobutadiene[54]

The formal combination of the levels RED and OX of general type A, n = 1, will produce the dicationic four-membered ring 44 which actually exists as a completely

symmetrical π-system. If X stands for a 4-pyridylrest with a N-carbethoxy group, the deep blue dication $45_{OX}c_{/RED}a$ results. Its UV-spectrum resembles that of a cyanine dye. This dication acts as the central species of a five-membered redox system: It has to be designated as the form OX of a two step redox system which is reduced via $45_{SEM}c$ to the brick red [4]-radialene $45_{RED}c$. On the other hand the blue dication functions as $45_{RED}a$ as the reduced form of another two step redox system. Here a new type of cyclobutadiene, namely $45_{OX}a$ is formed reversibly via the radical trication $45_{SEM}a$. Whereas only small energy differences are observed between $45_{RED}c$ and $45_{OX}c_{RED}a$ (19% $45_{SEM}c$ in equilibrium), $45_{OX}c_{/RED}a$ and $45_{OX}a$ are separated by a large energy barrier (>99,9% $45_{SEM}a$ in equilibrium). Since in $45_{OX}c_{/RED}a$ as well as in $45_{SEM}a$ the π-electrons are highly and symmetrically delocalized, localisation of the π-electrons in the different ring systems occur in the last step only. This step therefore includes an extra amount of antiaromatisation energy of ~50 kJ/mole in good agreement with other cyclobutadienes[55].

7.2.3 Electron Transfer Between a 1,3-Bismethylene Cyclobutane and a [1.1.0]-Bicyclobutane[56]

If one connects the two double bonded end groups X in form RED of general structure A (n = 1) with two methylene bridges instead of the usual σ-bond, then 46_{RED} should result. With well adapted X-groups this cyclobutane derivative 46_{RED} may

yield the bicyclobutane 46_{OX} on oxidation. In 46_{OX} the bridge, for which sp^{18}-hybridisation has been derived[57], plays the role of π-electrons in the general structure A_{OX}.

H3C   C6H5

CH3—N   N—CH3

H5C6   CH3

$\underline{47}$RED +e ↿⇂ -e

H3C   C6H5

CH3—N⊕   N⊕—CH3

H5C6   CH3

-e ↿⇂ +e $\underline{47}$OX

$$\left[CH_3-N \quad \quad N-CH_3 \right]^{\oplus}$$

H3C   C6H5

CH3—N   N—CH3

H5C6   CH3

T⇌

$$\left[CH_3-N \quad \quad N-CH_3 \right]^{\oplus}$$

H3C   C6H5

CH3—N   N—CH3

H5C6   CH3

$\underline{47}$ SEM₁

Single electron
in LUMO of RED

$\underline{47}$ SEM₂

Single electron
in HOMO of OX

With *47* such a reversible redox system has been realized. From electrochemical data again a two step electron transfer has to be derived although the concentration of *47*$_{SEM}$ is estimated to be as small as $10^{-3}\%$.

The difference in geometry of *47*$_{RED}$ (planar) and *47*$_{OX}$ (folded) together with the large difference in distance between C_1 and C_3 of the two rings automatically leads to different structures SEM$_1$ and SEM$_2$ of the intermediate oxidation form. If no high vibronic levels are involved these two radical ions may well be valence tautomeres as depicted in the redox scheme. System *47* resembles the dipyridylethylene system *5* in many respects[56].

7.3 Isocyclic Six-membered Rings as Bridges

Here and in Chap. 8.3 only those examples which are related to other systems of this report have been selected from a rather extensive investigation[58].

Logically, by insertion of a 1,4-phenylene group into the methine chain as in *48* the system is easily synthesized on the level *48*$_{RED}$. Transformation of *48*$_{RED}$ into *48*$_{OX}$ now demands some extra energy because a quino-dimethane bridge has to be produced at the same time, although two benzothiazolium units are formed.

This special situation probably causes the unexpectedly large K$_{SEM}$ of 10^3 which is associated with a rather persistent bluegreen radical ion *48*$_{SEM}$. Support of this interpretation is given by *49*$_{RED}$ in which the aromatic bridge is substituted by two vinylene groups of similar and fixed geometry. Here the difference E_2-E_1 becomes so small that only $E_2-E_1/2 = E_m$, which is shifted by 180 mV towards more negative potentials compared to *48*, can be determined. *49* itself behaves "normally". Its electrochemical data correspond to those of the benzothiazole derivative with three vinylene groups, *27* (n = 3) (5.1). Introduction of two more methine groups transforms *48* into *50*, which is now isolated as *50*$_{OX}$. Since on reduction the resonance energy of the phenylene bridge as well as that of the benzothiazolium rings has to be compensated, E_m of *50* is found 120 mV more negative than E_m of *49*, i.e. *50* needs higher energy for reduction than *49*. K_{SEM} of *50* stays small because the whole system has been extended compared to *48* and *49* (cf. 5.1.).

The effect of azasubstitution as in *48*$_{RED}$ $\longrightarrow$ *51*$_{RED}$ and *50*$_{OX}$ $\longrightarrow$ *52*$_{OX}$ is quite informative. In accordance with earlier observations (cf. 6.1.) the redox potentials become more positive by 0.8—0.9 V, however K_{SEM}'s are even diminished instead of being increased.

Table 10. Potentials E_1, E_2, E_m(V) and K_{SEM} versus Ag/AgCl in sat. KCl[58)]

Nr.	E_1	E_2	E_m	K_{SEM}	Solv.
48$_{RED}$	−0.04	+0.16	+0.06	$3 \cdot 10^3$	AN
49$_{RED}$	–	–	−0.12	~6	DMF
50$_{OX}$	–	–	−0.24	<4	DMA
51$_{RED}$	–	–	+0.83	18	AN
52$_{OX}$	–	–	+0.72	<5	DMF
53$_{OX}$	–	–	−0.34	21	AN
54$_{RED}$	+0.02	+0.11	+0.06	$1 \cdot 10^2$	AN

7.4 Heterocyclic Five-membered Rings as Bridges[58]

Formulas *53* and *54* depict two redox systems with a central thiophene group. For reasons already discussed they are isolated as *53*$_{OX}$ and *54*$_{RED}$. In spite of a longer

$\underline{53}_{OX}$ $\underline{54}_{RED}$

π-system in *54* its K_{SEM} is five times larger than that of *53* (Table 10). This irregularity probably has the same reason as in *48*. The somewhat lower K_{SEM} of *54* points to a more pronounced "butadiene character" of the thiophene link.

8 Redox Systems Containing Mono- or Non-quaternized Heterocycles[59, 60]

8.1 General Aspects

As already emphasized two step redox systems should exist with all compounds which can be derived from the general structures *A*, *B* and *C* (cf. 1.). These are isoelectronic and differ by their charges only.

From this point of view a rather informative comparison should be possible between the bisquaternary salts (OX) discussed so far (*Weitz* type, general structure *A*), the corresponding monoquaternary salts (OX) (general structure *B*) and the free bases (OX) (general structure *C*). These three systems have the advantage of being equivalent with regard to the extent and geometry of the π-systems. Again, this kind of comparison has to be based on reversible potentials since irreversible potentials can be used for estimation of E_1 and E_2 with special precautions only[61].

So far comparisons of isoelectronic systems of types *A*, *B* and *C* are restricted to few examples, because especially in type *B* the kinetic instability of the neutral radicals (SEM) becomes so pronounced that the level RED$^{\ominus}$ very often cannot be generated. This low persistency of *B* may be due to the fact that the odd electron is no longer distributed symmetrically over the molecule. In some cases dimeric products have been identified[62]. There are, however, tailor made radicals B_{SEM}, e.g. *55*, which exist as highly coloured, distillable compounds[63]. Nonetheless, from the

$\underline{55}$

Table 11. Comparison of semiquinone formation constants K_{SEM} of iso-π-electronic redox systems of the general types A, B and C in DMF versus Ag/AgCl in acetonitrile

		General Structure		
		A	B	C
	R^1	CH_3	CH_3	–
	R^2	CH_3 $(=1f)$	–	–
56		$2.8 \cdot 10^7$	$2.5 \cdot 10^{12}$	$3.7 \cdot 10^{11}$
57		$8 \cdot 10^{13}$	$3.7 \cdot 10^{15}$	$3.7 \cdot 10^{14}$
58		$2.4 \cdot 10^{12}$	$4.1 \cdot 10^{21}$	$1.6 \cdot 10^{15}$

following examples *56*, *57* and *58*, of which *57* exceeds the *Weitz* type[1], some general conclusions can be drawn (Table 11). In all three cases the thermodynamic stability of the radicals increases in the sequence $SEM_A^{\oplus} < SEM_C^{\ominus} < SEM_B$, in which the most sensitive SEM_B exhibits the highest equilibrium concentration. For this unexpected result a conclusive explanation is still missing. The sequence $K_{SEM}^A < K_{SEM}^C$ is already known from different (aromatic) hydrocarbons[64]. The difference in solvation energy between the radical cation and anion seems to be generally unimportant[65].

Table 12. Potentials E_1 and E_2 (V) of the iso-π-electronic systems *56A*, *56B* and *56C* in DMF versus Ag/AgCl in acetonitrile[59, 60]

			56 $R^1-N{=}{=}N-R^2$			
	C		B		A	
R^1	–		CH_3		CH_3	
R^2	–		–		CH_3	
$E_1(C)$	$E_2(C)$	$E_1(B)$	$E_2(B)$	$E_1(A)$	$E_2(A)$
$RED^{2\ominus}/SEM^{\ominus}$	$SEM^{\ominus}/OX$	$RED^{\ominus}/SEM$	$SEM/OX^{\oplus}$	$RED/SEM^{\oplus}$	$SEM^{\oplus}/OX^{2\oplus}$
-2.41	-1.76	-1.61	-0.82	-0.70	-0.33

As to be expected, the position of the potentials E_1 and E_2 of the iso-π-electronic systems are mainly charge controlled: With increasing positive charges the potentials are shifted to the more positive end. 4,4'-Bipyridyl, substituted to fit the general structures 56A, 56B and 56C, is a typical example for this influence of charges. In oxidation levels with equal charges electron transfer occurs nearly at the same potentials, e.g. $E_2(C) \sim E_1(B)$ and $E_2(B) \sim E_1(A)$ (Table 12). In 56C and other free bases of type C E_1 is located at so strongly negative potentials that only solvents from which impurities, especially protons, have been removed, will give reliable data (see later).

8.2 Comparison of Cationic and Anionic Iso-π-electronic Redox Systems[59, 60]

By comparison of about thirty pairs of isoelectric redox systems, namely bis-quaternary salts (potentials E_1^c and E_2^c (e.g. 56A) and the corresponding free bases (potentials E_1^a and E_2^a (e.g. 56C)) a general relationship between their redox properties has been discovered. It turns out that potentials E_1^c and E_1^a as well as E_2^c and E_2^a are related to the distance between the nitrogen atoms by the linear Eq. (6)

$$E_{1(2)}^c - E_{1(2)}^a = A + B \cdot \frac{1}{d} \tag{6}$$

in which A and B represent constants and d the distance between the nitrogen atoms in the specific system, compared to 4,4'-bipyridyl which is taken as a standard with $d = 1$.

Interestingly, d is not relevant for the difference between E_2^a (SEM$^\ominus$/OX) and E_1^c (RED/SEM$^\oplus$) which is always around 1.0V. In these cases one member of the redox pair is always uncharged. In contrast both differences $E_2^a - E_1^a$ and $E_2^c - E_1^c$ depend on the forementioned distances according to $^1/d \cdot 0.4$ V. Here the transfered electron is strongly influenced by the charges already present. Since positive and negative charges exhibit similar effects on the potentials measured in acetonitrile or DMF, ion pairing seems to be absent in both pairs of systems. Figure 10 demonstrates this correlation which allows estimation of the three remaining potentials if one is known.

8.3 Vinylogous Redox Systems Containing Non-quaternized Heterocycles

Under carefully controlled conditions K_{SEM}'s of the vinylogous bases $59_{OX} - 63_{OX}$, which are collected in Table 13, can be derived from their potentials E_1 and E_2. Com-

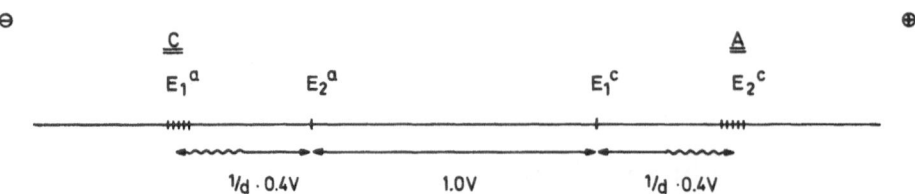

Fig. 10. Correlation between potentials E_1^a, E_2^a of free bases (type C) and of their corresponding, iso-π-electronic bisquaternary salts, E_1^c, E_2^c (type A)[59, 60]

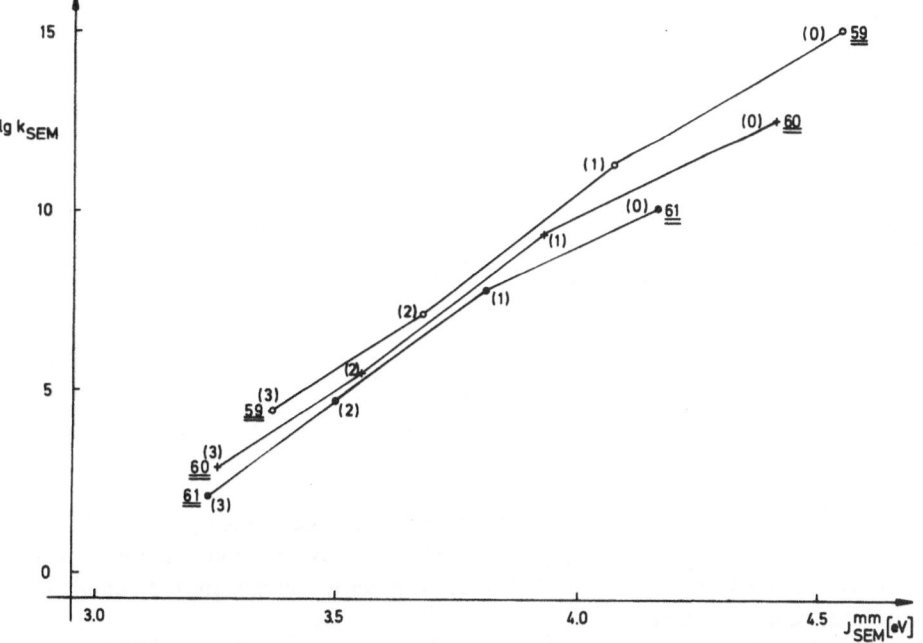

$\underline{\underline{59}}$OX $\underline{\underline{60}}$OX $\underline{\underline{61}}$OX

$\underline{\underline{62}}$OX $\underline{\underline{63}}$OX

Table 13. K_{SEM}'s, calculated from E_1 and E_2 of the vinylogous bases $59_{OX}-63_{OX}$ in DMF versus Ag/AgCl in acetonitrile at 25 °C

	n = 0	n = 1	n = 2	n = 3
59_{OX}	$8.2 \cdot 10^{10}$	$4.1 \cdot 10^8$	$3.3 \cdot 10^5$	$3.6 \cdot 10^3$
60_{OX}	$1.1 \cdot 10^{11}$	$7.3 \cdot 10^7$	$5.4 \cdot 10^4$	$4.9 \cdot 10^2$
61_{OX}	$2.2 \cdot 10^6$	$1.5 \cdot 10^6$	$6.8 \cdot 10^3$	$6.6 \cdot 10^1$
62_{OX}	$3.9 \cdot 10^3$	$6.0 \cdot 10^4$	$4.1 \cdot 10^2$	$1.0 \cdot 10^1$
63_{OX}	$3.3 \cdot 10^{10}$	$2.9 \cdot 10^7$	$5.6 \cdot 10^4$	$2.3 \cdot 10^2$

Fig. 11. Correlation between lg K_{SEM} of vinylogous redox systems $59_{OX}-61_{OX}$ and the Coulomb repulsion integral J_{mm}^{SEM} [38, 66)

33

pared to the corresponding cationic redox systems (5.1., Table 8) K_{SEM}'s of the anionic systems are larger by a factor of $10^3 - 10^6$! Nevertheless, these K_{SEM}'s also decrease drastically with increasing chain length. Again, as had already been shown for diphenylpolyenes[15], there exists a linear correlation between lg K_{SEM} and the Coulomb repulsion integral J_{mm}^{SEM} (Fig. 11), thus underlining the generality of the abovementioned relation (cf. 5.1). Deviations to smaller K_{SEM}'s clearly arise from nonplanarity of the system. Curiously the calculations fit best if the same nitrogen parameters are used in both cases[38].

8.4 Acetylene Bridged Non-quaternized N-Heterocycles

By comparison of the already discussed redox systems $59_{OX} - 63_{OX}$ with $64_{OX} - 70_{OX}$[38, 67] the different effects of ethylene and acetylene bridges can be evaluated.

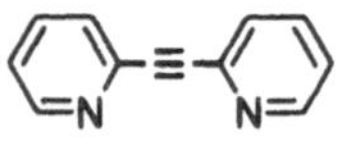

64OX

$K_{SEM} = 6.8 \cdot 10^{10}$

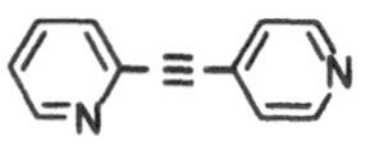

65OX

$K_{SEM} = 3.8 \cdot 10^5$

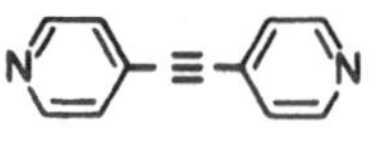

66OX

$K_{SEM} = 7.4 \cdot 10^7$

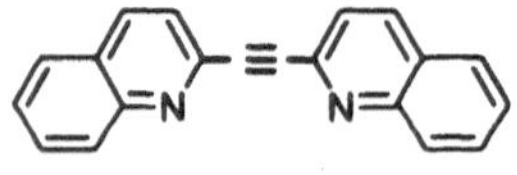

67OX

$K_{SEM} = 6.9 \cdot 10^8$

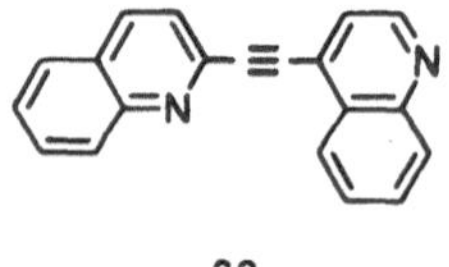

68OX

$K_{SEM} = 1.2 \cdot 10^9$

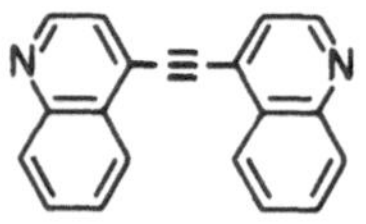

69OX

$K_{SEM} = 9.3 \cdot 10^6$

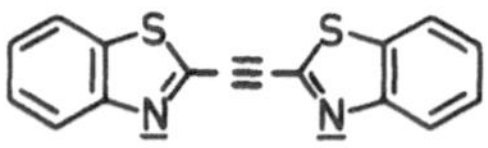

70OX

$K_{SEM} = 8.9 \cdot 10^6$

For $64-70$ reversible potentials RED$^{2\ominus}$/SEM$^{\ominus}$ can be obtained at $-55\ °C$ only in DMF from which protic impurities have been removed[68]. One should keep in mind that here the levels RED (and SEM) exist as heterocyclic substituted cumulenes which may rapidly polymerize.

The effect of the extra π-bond in $64_{OX} - 70_{OX}$ is remarkably small. The rather high K_{SEM}'s as well as the absolute positions of E_1 and E_2 correspond to those of the ethylene bridged bases (cf. 8.3). The same behaviour has been reported for the pair dibenzoylethylene/dibenzoylacetylene[69].

For unknown reasons the sequences of K_{SEM}'s with respect to the linking positions of the heterocycles pyridine and quinoline are not the same in both series. K_{SEM} of 65, however, had to be derived from a poorly reversible potential E_1.

A direct comparison of the bisquaternized isomeric dipyridyl ethylenes (cf. 2.3.) with the corresponding bisquaternary acetylenic bases is still lacking, since so far the latter could not be reversibly reduced.

9 Medium Effects

9.1 General Remarks

The effect of the reaction medium on the three levels RED, SEM and OX on redox systems of the general types *A*, *B* and *C* can be manifold. Since two of three oxidation levels carry charges solvation energies will affect not only the absolute positions of E_1 and E_2 but also ΔE and thereby K_{SEM}. Only if solvation energies are similar, comparison of different redox systems will reflect structural properties. This seems to be the case with the *Weitz* types under discussion in DMF and acetonitrile[7]. Complications may, however, arise from larger aggregates, especially ion pairs in solvents of low polarity[22]. More often rapid reversible or irreversible consecutive reactions may disturb the equilibrium between the three oxidation levels[1]. In principle RED is always sensitive to electrophiles (especially $RED^{2\ominus}$ at very negative potentials), OX, however, to nucleophiles (especially $OX^{2\ominus}$ at very positive potentials). The intermediate oxidation level SEM may be attacked both by nucleophiles and electrophiles, depending on its charge and the position of the potentials. Besides, SEM can react as a radical. This flexible reactivity is also found with *Wurster's* radical ions[70].

A highly efficient reagent for removing electrophilic and nucleophilic impurities is neutral aluminium oxide, which is used immediately before preparation of the solution[71] or during the voltammetric measurement[68].

Because of the extremely high mobility of the redox equilibria (see Fig. 13) rapid consecutive reaction of just one member of the system will disturb these equilibria. For the same reason kinetic stability is the crucial property for detection and isolation of one or more components of the redox system. E. g. 32_{SEM} can be easily determined by UV/VIS- and ESR-spectroscopy in spite of K_{SEM} being as small as 0.003. In contrast $21B$ with K_{SEM} $6 \cdot 10^{14}$ and the corresponding system with two $(CH_2)_3$-bridges (K_{SEM} $6 \cdot 10^5$) could not be characterized by their UV/VIS-spectra because they decay too rapidly.

Unfortunately in quite a few cases voltammetric data become inaccessible because the electroactive compound is adsorbed on the electrode. By variation of the electrode material, the concentration of the different solutes, the solvent and the scanning time adsorption phenomena may occasionally be overcome.

The most important solvent effects will be discussed in the following sections.

9.2 Protonation Reactions

In the already discussed diaza-substituted redox systems, which are represented by the general formula 71, the reduced level 71_{RED} is not only an azine but also contains two amidrazone units. It is therefore understandable that 71_{RED} is easily stepwise protonated. (About the less important protonation of 71_{SEM} cf. 1.c.[89] and Fig. 13.)

In a solution containing more than 99.9% of violet to blue 71_{SEM} (K_{SEM} $10^7 - 10^{10}$ cf. 6.1.) a yellow to red colour is immediately produced by excess acid. With base the colour of 71_{SEM} is also immediately restored, a phenomenon which resembles that of a normal acid-base indicator. UV-spectra, however, disclose an acid-driven disproportionation which transforms 71_{SEM} quantitatively into 71_{OX} and and $71_{RED} H^{\oplus}$ or $71_{RED} 2H^{\oplus}$ respectively. If the free azine 71_{RED} is reformed by base it comproportionates[72] with 71_{OX}, building up the original concentration of 71_{SEM}.

Redox potentials of species sensitive to protons are pH-dependent according to the following Eq. (7)[73].

$$\frac{dE}{dpH} = -2.30 \frac{RT}{F} \cdot \frac{a}{b} \frac{V}{pH} = -59 \frac{a}{b} \frac{mV}{pH} \text{ at } 25\,^{\circ}C \tag{7}$$

(a = difference in the number of protons, b = difference in the number of electrons between higher and lower oxidation levels).

Therefore the forementioned diaza-systems of type 71 show pH-dependent redox behaviour which, however, is restricted to E_1(RED/SEM). Its pH-dependence amounts to 58–60 mV/pH-unit, signalizing the equilibrium $71_{RED} + H^{\oplus} \rightleftharpoons 71_{RED} H^{\oplus}$. For the same reason lg K_{SEM} is simultaneously diminished by one unit if pH is lowered by one unit. Figure 12 demonstrates this pH-dependence for the already discussed systems $33, 35, 72$ and 73.

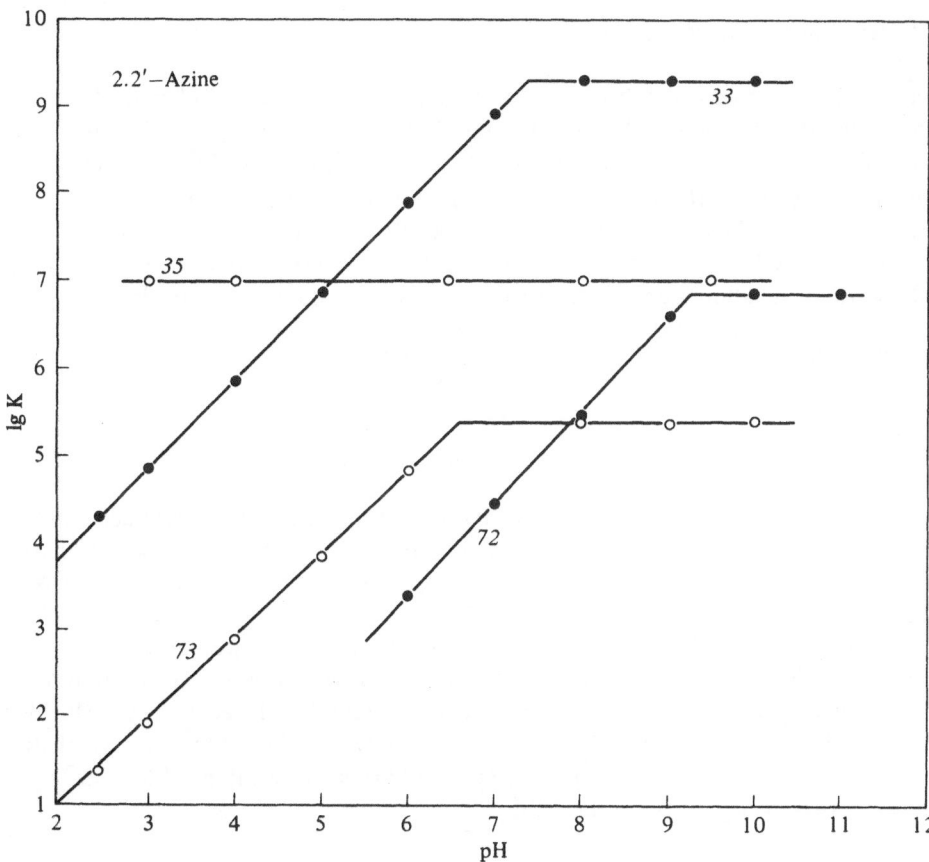

Fig. 12. Correlation of lg K_{SEM} with pH in water/methoxy ethanol 1:1 for diaza redox systems *33, 35, 72* and *73* (N-ethylgroups instead of N-methylgroups)[74)]

__72__ RED   __73__ RED

The intersections of the pH-dependent with the pH-independent parts of the curves in Fig. 12 mark the pK-value of the corresponding azine. With the very weakly basic *35* no protonation occurs even at pH 2.9.

As already mentioned, dipyridinium salts *1*, together with $5-7$[6)] cannot reversibly be reduced down in RED in water. Due to its high basicity, RED is probably protonated, whereby e.g. *74* may be formed, which as a dienamine could be consumed by consecutive reactions.

$$CH_3-N \quad \quad N-CH_3 \quad \overset{+H}{\underset{-H}{\rightleftharpoons}} \quad CH_3-\overset{\oplus}{N} \quad \quad N-CH_3$$

__1f__ RED   __74__

9.3 Reactions with Nucleophiles

Even carefully purified acetonitrile and DMF still contain nucleophilic impurities, which can interfer with the electron transfer SEM/OX so strongly that it becomes irreversible and E_2 cannot be derived from the polarographic data. With E_2's rather positive, as e.g. in *15*, this behaviour is most pronounced. Probably the adduct of *15*$_{OX}$ and nucleophiles e. g. *75* undergoes consecutive reactions like ring openings. In these cases addition of some perchloric acid[28, 75], trifluoroacetic acid or its anhydride[76] and aluminium oxide[68] will completely suppress such interfering

15$_{OX}$ *75*

reactions. This interpretation is supported by the fact that the influence of nucleophilic impurities can be overcome by fast cyclic voltammetry (1—20 V/s, i.e. low stationary concentration of OX) and by low temperatures ($\sim -40\ ^\circ$C, i. e. slow reaction of OX with nucleophiles).

For the same reasons the forms RED with relatively positive potentials seem to be stable to air in a pure state or a pure solution. In the presence of some acid, however, the colour of the radical cation SEM$^\oplus$ develops rapidly. The acid traps the highly nucleophilic $\cdot O_2^{\ominus}$[77] formed in the course of autoxidation. Thus SEM$^\oplus$ and especially OX$^{2\oplus}$ are no longer removed from the redox-equilibrium since anions of low nucleophilicity are provided by the added acid.

9.4 Equilibria of Association

All discussed K_{SEM}'s are related to the separately solvated members of a two step redox system. In solvents of low polarity the charged forms may form ion pairs. Especially prone to this association are anionic redox systems of Type C (RED$^{2\ominus}$ + OX $\rightleftarrows$ 2 SEM$^\ominus$) since the often used gegenions K$^\oplus$, Na$^\oplus$ and Li$^\oplus$ tend to form ion pairs with the anions. These exhibit special UV/VIS-, NMR- and ESR-spectra[78] as well as g-values[65, 79]. Dimeres of the type (SEM$^\ominus$M$^\oplus$)$_2$ may also be formed, as demonstrated with the anion radicals of pyrazines[80], heptafulvalene[81] and tetracyano-quinodimethanes[82]. Corresponding associations are reported for dianions derived e. g. from 1,4-dihydroxy naphthaline[83] and aromatic hydrocarbons[68]. Of course ion pair formation is increased with decreasing solvation power of the medium for the relatively small cations[78]. (Cf. the series of donor qualities e. g. CH$_2$Cl$_2$ < Et$_2$O < THF < CH$_3$CN < DMF < HMPT[84, 85].) Strong complexing ligands such as crownethers separate these ion pairs[65].

In cationic redox systems of type *A* with SEM$^\oplus$ and OX$^{2\oplus}$ ion pairing can be more easily avoided by choosing anions as BF$_4^\ominus$, ClO$_4^\ominus$, PF$_6^\ominus$ and B(C$_6$H$_5$)$_4^\ominus$ of low nucleophilicities and low solvation energies[86]. With these precautions $5 \cdot 10^{-4}$ molar solutions of type *A* redox systems in acetonitrile or DMF hardly show any anomalies, even in the presence of $2 \cdot 10^{-1}$ m Et$_4$N$^\oplus$BF$_4^\ominus$. In methylenechloride,

$$\underline{\underline{76}}_{SEM} \qquad\qquad (\underline{\underline{76}}_{SEM} BF_4)_2$$

however, 76_{SEM} dimerizes reversibly if its concentration is raised from 10^{-5} to 10^{-4} molar[22]. The dimer, which has lost the paramagnetism of 76_{SEM} shows a broad charge transfer band, whereas the typical absorptions of 76_{RED} and 76_{OX} are absent[22]. As a plausible structure $(76_{SEM} BF_4)_2$ is therefore proposed, in which the spins of the two radicals are antiparallel and the π-systems are much less affected by the counterions than in the corresponding monomeric ion pair.

Interestingly, the formation of diamagnetic dimers from radical cations has also been observed for 1_{SEM} with different R groups — even in aequous[62] or methanolic[87] solution, by raising their concentration or by lowering the temperature. The dimers again exhibit typical charge transfer absorption bands[62]. Since the halogenid ions used should be highly solvated in methanol dimerisation through ion pairing by charge attraction is less likely here. As to be seen from Table 14 the dimerisation of 1_{SEM} becomes more exothermic and the entropy less positive with increasing size of the

Table 14. Thermodynamic constants for the equilibrium $2\,1_{SEM} \rightleftharpoons (1_{SEM})_2$ in methanol at 25 °C [87]

R	$X^{\ominus}$	ΔH° kJ mol^{-1}	ΔG° kJ mol^{-1}	ΔS° J K^{-1} mol^{-1}
H	Br$^{\ominus}$	− 4.25	−16.16	+34.76
CH$_3^a$	Cl$^{\ominus}$	− 6.77	−16.37	+32.21
C$_2$H$_5$	I$^{\ominus}$	− 8.19	−15.94	+26.01
n-C$_3$H$_7$	I$^{\ominus}$	−10.46	−15.17	+15.80
n-C$_4$H$_9$	Cl$^{\ominus}$	−12.07	−15.15	+10.34
CH$_2$–C$_6$H$_5$	Br$^{\ominus}$	−22.97	−13.67	−31.19
$-CH_2\overset{O}{\overset{\|}{C}}-N$ (morpholino, CH$_3$)^b	Cl$^{\ominus}$	−45.05	−10.60	−115.6

^{a} Paraquat$^{®}$ ^{b} Morphamquat$^{®}$

R groups. The unexpected positive $\Delta S°$ values for smaller R groups point to a stronger solvent shell which has to be removed on dimerisation. Therefore, a face to face structure of the dimer similar to $(76_{SEM})_2$ has been proposed in accordance with conclusions from the UV-spectra of Paraquat® in aequous solution[62]. The stretched structure of 1_{SEM} seems to be important since the radical cation derived from Diquat® $(2B2_{SEM})$ does not dimerize[88].

10 Kinetics of Electron Transfer Reactions

All chemical and electrochemical experimental evidence of the redox systems under discussion points to a nearly instant electron transfer. The abovementioned azine 73_{RED} has been investigated very carefully with respect to its redox kinetics at different acidities[89]. As depicted in Fig. 13 even $RED\,H^\oplus$ and $RED\,H_2^{2\oplus}$ exhibit redox

Fig. 13. Electron and proton transfers with 73_{RED} in water/methoxyethanol 1:1 at 25 °C and ionic strength 0.5 (k in $mol^{-1}\,s^{-1}$)[89]

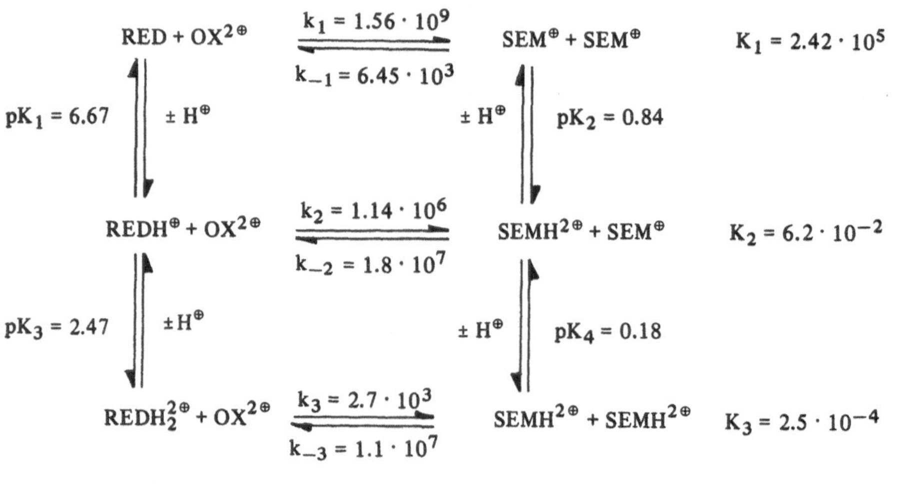

activity. The corresponding K_{SEM}'s, however, are smaller by factors of 10^7 and 10^9 respectively. In case of $RED\,H_2^{2\oplus}$ electron and proton transfer to $OX^{2\oplus}$ seems to occur simultaneously.

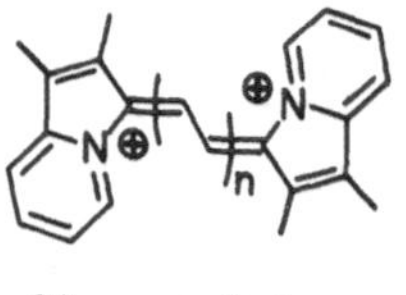

$\underline{30}_{OX}$ n=0-4

$k_1 = 6.5 \cdot 10^8\,mol^{-1}\,s^{-1}$
$k_2 = 8.1 \cdot 10^5\,mol^{-1}\,s^{-1}$
$K_{SEM} = 800$

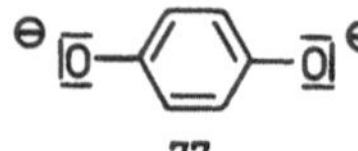

$\underline{77}$ RED

$k_1 = 2.6 \cdot 10^8\,mol^{-1}\,s^{-1}$
$k_2 = 6.8 \cdot 10^7\,mol^{-1}\,s^{-1}$
$K_{SEM} = 3.8$

All reaction constants are very large, even approaching diffusion controlled transfer rates[90], depending on K_{SEM}. Very probably 73_{RED} reflects the kinetic behaviour of all redox systems of the general types A, B and C. Another example of type A ($30^{91)}$) and one of type C ($77^{92)}$) exhibit similar properties.

11 Outlook

The remarkable variety of redox systems which can already be derived from the *Weitz* type underline the wide scope of the general structure A and C as a basic principle for two step redox systems. The empirical material as well as general rules regarding structural influences on potentials and K_{SEM}'s have been developed to such an extent, that redox systems can be taylored to meet special purposes. Catalysts for electron transfer, light sensitive systems and compounds of high electrical conductivity are some fields in which these redox systems could occupy key positions. Some applications have already been discussed in a previous review of wider scope[1])

Acknowledgements. In this report numerous results from our laboratory have been included, which have been brought to light through the enthusiastic and critical work of my former students and post-doctoral coworkers to whom I owe my gratitude: Heinz Balli, Horst Berneth, Horst Conrad, Klaus Deuchert, Werner Daum, Hans A. Dresch, Uwe Fritzsche, Brian J. Garner, Jörg Groß, Bernd Hagenbruch, Michael Horner, Günter Kießlich, Edouard Frederic Lier, Friedrich Linhart, Hermann Pütter, Helmut Quast, Günther Ruider, Wolfgang Schenk, Dieter Scheutzow, Peter Schilling, Helmut Schlaf, Albrecht Schott, Hans-Christian Steinmetzer, Ingo Stemmler.

Financial support of our research programme by Deutsche Forschungsgemeinschaft, Dechema and Fonds der Chemischen Industrie ist greatly acknowledged.

12 References

1. a) Deuchert, K., Hünig, S.: Angew. Chem. *90*, 927 (1978), Int. Ed. Engl. *17*, 875 (1978);
   b) Hünig, S. et al., Pure Appl. Chem. *15*, 109 (1967)
2. a) Dähne, S.: Z. Chem. *5*, 441 (1965); cf. Science *199*, 1163 (1978); b) Fabian, J., Hartmann, H.: Z. Chem. *13*, 261 (1973); J. Signal AM *2*, 457 (1974); Theoret. Chim. Acta *36*, 351 (1975); J. Mol. Structure *27*, 67 (1975)
3. Cf. also Hanson, P.: Heteroaromatic Radicals, Part I: General Properties; Radicals with Group V Ring Heteroatoms. In Adv. Heteroc. Chem. *25*, 205 (1979)
4. a) Michaelis, L.: Chem. Rev. *16*, 243 (1935)
   b) Michaelis, L., Hill, E. S.: J. Gen. Physiol. *16*, 859 (1933)
5. Weitz, E.: Angew. Chem. *66*, 658 (1954); Weitz, E., Fischer, K.: Angew. Chem. *38*, 1110 (1925)
6. a) Boon, R. W.: Chem. Ind. (London) *1965*, 782; b) Michaelis, L., Hill, E. S.: J. Gen. Physiol. *16*, 856 (1933); J. Amer. Chem. Soc. *55*, 1481 (1933); c) Homer, R. F., Tomlinson, T. E.: J. Chem. Soc. *1960*, 2498; d) Summers, L. A.: Nature (London) *214*, 381 (1967); e) Homer, R. F., Mees, G. C., Tomlinson, T. E.: J. Sci. Food Agr. *11*, 309 (1960)
7. Hünig, S., Groß, J., Schenk, W.: Liebigs Ann. Chem. *1973*, 324
8. Zuman, P.: Substituent Effects in Organic Polarography. New York: Plenum Press. 1967; cf. Hine, J.: Structural Effects on Equilibria in Organic Chemistry. New York: J. Wiley 1975

9. Taft, R. W., Jr.: Separation of Polar, Steric and Resonance Effects in Reactivity, published in Steric Effects in Organic Chemistry. Newman, M. S. (ed.). New York: J. Wiley 1965; Taft, R. W., Jr. et al.: J. Am. Chem. Soc. *85*, 709 (1963)

10. Hünig, S., Schenk, W.: Liebigs Ann. Chem. *1979*, 1523

11. Horner, M.: Dissertation Würzburg 1977

12. Cf. l. c.[9]. Absorption spectra of some of these derivatives have been reported earlier:
    a) Kosower, E. M., Cotter, J. L.: J. Am. Chem. Soc. *86*, 5524 (1964);
    b) Bruin, F. et al.: J. Chem. Phys. *36*, 2783 (1962); Nielsen, T., Moore, W., Berry, M. M.: J. Org. Chem. *29*, 2175 (1964)

13. a) Summers, L. A.: Naturwissenschaften *54*, 491 (1967); b) Cf. Bartle, K. D., Jones, D. W.: Adv. Organ. Chem., Vol. 8, p. 317. Taylor, E. C. (ed.) New York: J. Wiley 1972 New York: J. Wiley 1972

14. A diheral angle of 20° was determined by X-ray analysis: Derry, J. E., Hamor, T. E.: Nature (London) *221*, 464 (1969); Sullivan, P. D., Williams, M. L.: J. Am. Chem. Soc. *98*, 1711 (1976)

15. Čarsky, P. et al.: Tetrahedron *25*, 4781 (1969)

16. Hünig, S. et al.: Liebigs Ann. Chem. *1973*, 339

17. Zweig, A., Maurer, A. H., Roberts, B. G : J. Org. Chem. *32*, 1322 (1967)

18. Pointeau, R.: Ann. Chim. *7*, 669 (1962)

19. Bruhin, J., Gerson, F.: Helv. Chim. Acta *58*, 2422 (1975)

20. Berneth, H., Hünig, S.: 1979, unpublished

21. Hünig, S., Dresch, H. A., Horner, M.: 1975, unpublished

22. Hünig, S. et al.: Liebigs Ann. Chem. *1973*, 1036

23. a) Hünig, S., Ruider, G., Scheutzow, D.: 1967, unpublished; b) Yoshida, Z., Sagimoto, T., Yoneda, S.: J. Chem. Soc. Chem. Comm. 1972, 60

24. a) Johnson, C. S., Jr., Gutowsky, H. S.: J. Chem. Phys. *39*, 58 (1963); b) Grossi, L., Minisci, F., Pedulli, G. F.: J. Chem. Soc. Perkin Trans. II *1977*, 943

25. Perlstein, J. H.: Angew. Chem. Int. Ed. Engl. *16*, 519 (1977); Z. G. Soos: J. Chem. Ed. *55*, 546 (1978); Torrance, J. B.: Acc. Chem. Res. *12*, 79 (1979)

26. Coffen, D. L. et al.: J. Am. Chem. Soc. *93*, 2258 (1971)

27. Engler, E. M. et al.: J. Am. Chem. Soc. *97*, 2921 (1975)

28. a) Hünig, S. et al.: Liebigs Ann. Chem. *1973*, 310; b) Hünig, S. et al.: Int. J. Sulfur Chem. C, *6* (1971) 109

29. Wheland, R. C., Gillson, J. L.: J. Am. Chem. Soc. *98*, 3916 (1976)

30. Green, D. C.: J. C. S. Chem. Comm. *1977*, 161

31. Scott, B. A., Kaufman, F. B., Engler, E. M.: J. Am. Chem. Soc. *98*, 4342 (1976)

32. Schukath, G., Le Van Hinh, Fanghänel, E.: Z. Chem. *16*, 360 (1976)

33. Engler, E. M. et al.: J. Am. Chem. Soc. *100*, 3769 (1978)

34. Miles, M. G. et al.: J. Chem. Soc. Chem. Comm. *1974*, 751

35. Hurtley, W. R. H., Smiles, S.: J. Chem. Soc. *1926*, 1821, 2263; Electronic structure of *18*RED cf. J. Spanget-Larsen, R. Gleiter, S. Hünig: Chem. Phys. Lett. *37*, 1, 29 (1976)

36. Hünig, S., Scheutzow, D., Schlaf, H.: Liebigs Ann. Chem. *765*, 126 (1972)

37. Hünig, S., Steinmetzer, H. Ch.: Liebigs Ann. Chem. *1976*, 1060

38. Stemmler, I.: Dissert. Univers. Würzburg, 1976

39. Hünig, S. et al.: Liebigs Ann. Chem. *1974*, 1436

40. Hünig, S., Linhart, F.: Liebigs Ann. Chem. *1976*, 317

41. Hünig, S. et al.: J. Phys. Chem. *75*, 335 (1971)

42. Čarsky, P. et al.: Collect. Czech. Chem. Commun. *36*, 560 (1971)

43. Hünig, S., Linhart, F., Scheutzow, D.: Liebigs Ann. Chem. *1975*, 2102

44. Hünig, S., Linhart, F.: Liebigs Ann. Chem. *1975*, 2116

45. Michaelis, L., Schubert, M. P., Granich, S.: J. Am. Chem. Soc. *61*, 1981 (1939)

46. a) Kuhn, H.: Helv. Chim. Acta *31*, 1441 (1948); b) Kuhn, H.: J. Chem. Physics *16*, 840 (1948); *17*, 1198 (1949); Z. Elektrochem. *53*, 165 (1949); c) Simpson, W. T.: J. Chem. Physics *16*, 1124 (1948); d) Bayliss, N. S.: Quart. Rev. (Chem. Soc. London) *6*, 319 (1952); e) Kuhn, H.: Angew. Chem. *71*, 93 (1959); f) Kuhn, H. in: Optische Anregung Organischer

Systeme, p. 70/71. Weinheim: Verlag Chemie 1966; see also Fortschr. org. Naturstoffe *16*, 169 (1958); *17*, 404 (1959)

47. König, W.: J. Prakt. Chem. *112* (2), 1 (1926)

48. In the cyanine series similar effects are observed; cf. l. c.[49]

49. Sturmer, D. W. in: Spec. Top. Heterocyclic Chem. p. 441. Weissberger, A. and Taylor, E. C. (ed.) New York: J. Wiley 1977

50. Hünig, S. et al.: Liebigs Ann. Chem. *676*, 36 (1964)

51. Hünig, S. et al.: Liebigs Ann. Chem. *1974*, 1423

52. Berneth, H.: Dissert. Univers. Würzburg, 1978

53. Rieke, R. D., Rich, W. E., Ridgeway, J. H.: J. Am. Chem. Soc. *93*, 1962 (1971)

54. Horner, M., Hünig, S.: Angew. Chem. *89*, 424 (1977)

55. Breslow, R. et al.: J. Am. Chem. Soc. *95*, 6688 (1973)

56. a) Horner, M., Hünig, S.: J. Am. Chem. Soc. *99*, 6120 (1977); b) Horner, M., Hünig, S.: J. Am. Chem. Soc. *99*, 6122 (1977)

57. a) Pomerantz, M., Abrahamson, E. W.: J. Am. Chem. Soc. *88*, 3970 (1966); b) Newton, M., D., Schulman, J. M.: ibid., *94*, 767 (1972); c) Newton, M. D., Schulman, J. M.: J. Am. Chem. Soc., *96*, 6295 (1974); d) Schulman, J. M., Venanzi, T. J.: Tetrahedron Lett. 1461 (1976); e) Pomerantz, M., Fink, R., Gray, G. A.: J. Am. Chem. Soc. *98*, 291 (1976)

58. Fleckenstein, E.: Dissert. Univers. Würzburg 1969

59. Hünig, S., Pütter, H.: 1972, unpublished

60. Hünig, S., Horner, M.: 1977, unpublished

61. Cf. Streitwieser, A. Jr., Schwager, I.: J. Phys. Chem. *66*, 2316 (1962)

62. Kosower, E. M., Coffer, J. L.: J. Am. Chem. Soc. *86*, 5524 (1964)

63. Kosower, E. M., Poziomek, E. J.: J. Am. Chem. Soc. *86*, 5515 (1964)

64. Cf. l. c. 1) compounds *15–18* versus *19–22* (R = H); Diphenylanthracene, reduction: Jensen, B. S., Parker, V. D.: J. Am. Chem. Soc. *97*, 5211 (1975), Oxidation: l. c. [76]

65. Hogen-Esch, T. E.: Adv. Phys. Org. Chem., Vol. 15, p. 153. Gold, V. and Bethell, D. (ed.) 1977

66. K_{SEM}'s are taken from potentials at $-55\,°C$, because they are more reliable especially for members n = 0

67. Fritsche, U.: Dissert. Univers. Würzburg, 1973

68. a) Jensen, B. S., Parker, V. D.: J. Am. Chem. Soc. *97*, 5211 (1975); b) Jensen, B. S., Parker, V. D.: J. Chem. Soc. Chem. Commun. *1974*, 367; c) Hammerich, D., Parker, V. D.: Electrochim. Acta *18*, 537 (1973)

69. Breslow, R. et al.: J. Am. Chem. Soc. *96*, 249 (1974)

70. Hünig, S., Daum, W.: Liebigs Ann. Chem. *595*, 131 (1955)

71. Lines, R., Jensen, B. S., Parker, V. D.: Acta Chem. Scand. *B32*, 510 (1978)

72. This expression was introduced by Anschütz, L., Broeker, K., Ohneiser, A.: Ber. dtsch. chem. Ges. *77*, 443 (1944)

73. Clark, W. M., Cohen, B.: Public Health Rep. *38*, 666 (1923); Clark, W. M.: ibid. *38*, 443 (1923), Chem. Rev. *2*, 127 (1926)

74. Hünig, S. et al.: Liebigs Ann. Chem. *676*, 52 (1964)

75. Hünig, S. et al.: Liebigs Ann. Chem. *752*, 196 (1971)

76. Hammerich, O., Parker, V. D.: J. Am. Chem. Soc. *96*, 4289 (1974)

77. Lissel, M., Dehmlow, E.: Tetrahedron Lett. *1978*, 3689

78. Smid, J.: Angew. Chem. *84*, 127 (1972), Int. Ed. Engl. *11*, 112 (1972)

79. Stevenson, G. R., Alegria, A. E., Block, A. Mc. B.: J. Am. Chem. Soc. *97*, 4859 (1975)

80. Hag, R. S., Pomery, P. J.: Austr. J. Chem. *24*, 2287 (1971)

81. Sevilla, M. D. et al.: J. Am. Chem. Soc. *91*, 4139 (1969)

82. Boyd, R. H., Philips, U. D.: J. Chem. Phys. *43*, 2927 (1965)

83. Jaworski, J. S., Kalinowski, M. K.: Rosz. Chem. *49*, 1141 (1975)

84. Normant, H.: Angew. Chem. *79*, 1029 (1967), Int. Ed. Engl. *6*, 1046 (1967)

85. Reichardt, Ch.: Angew. Chem. *91*, 119 (1979), Int. Ed. Engl. *18*, 98 (1979); Reichardt, Ch.: Solvent Effects in Organic Chemistry. Weinheim, New York: Verlag Chemie 1979

86. Parker, A. J.: Chem. Rev. *69* (1), 1 (1969)

S. Hünig and H. Berneth

87. Evans, E. G., Evans, J. G., Baker, M. W.: J. Chem. Soc. Perkin Trans. II, *1977*, 1787
88. Evans, E. G., Alford, R. E., Rees, N. H.: J. Chem. Soc. Perkin Trans. II, *1977*, 445
89. Bernasconi, C. F., Bergström, R. G., Boyle, W. J., Jr.: J. Am. Chem. Soc. *96*, 4643 (1974)
90. Eigen, M.: Angew. Chem. *75*, 489 (1963), Int. Ed. Engl. *3*, 1 (1964)
91. Bennion, B. C., Auborn, J. J., Eyrig, E. M.: J. Phys. Chem. *76*, 701 (1972)
92. Bernasconi, C. F., Wang, H.: J. Am. Chem. Soc. *99*, 2214 (1977)

Received August 10, 1979

Controlling Factors in Homogeneous Transition-Metal Catalysis
Control of Metal-Catalyzed Reactions, Part XII*, **

Paul Heimbach and Hartmut Schenkluhn

Universität Essen, Institut für Organische Chemie, Postfach 6843, D-4300 Essen 1, Federal Republic of Germany
and
Max-Planck-Institut für Kohlenforschung, Postfach 011325, D-4330 Mülheim 1, Federal Republic of Germany

Table of Contents

---

*   Part XI: see Ref. [115]

** Contributions from: N. Arfsten (Mülheim), H. Bandmann (Essen), R. Berger (Mülheim), F. Brille (Essen), B. Gerding (Essen), P. Heimbach (Essen), E. Hübinger (Mülheim), J. Kluth (Mülheim/Essen), H. Krische (Mülheim), E. F. Nabbefeld-Arnold (Mülheim), B. Pittel (Essen), A. Roloff (Mülheim), W. Scheidt (Mülheim), H. Schenkluhn (Essen), A. Sisak (Mülheim), B. Weimann (Mülheim), M. Zähres (Essen)

1 Introduction

Ni-organic chemistry of butadiene started in the early 1950's when this substrate could be slowly dimerized to cis, cis-cycloocta-1,5-diene[1] in relatively low yields (~40%) by using so-called Reppe catalysts of the type $L_2Ni(CO)_2$. Carbonyl-free low-valent Ni-complexes and catalysts on the basis of Cr and Ti were more effective and also led to further ring-syntheses (Scheme 1-1)[2].

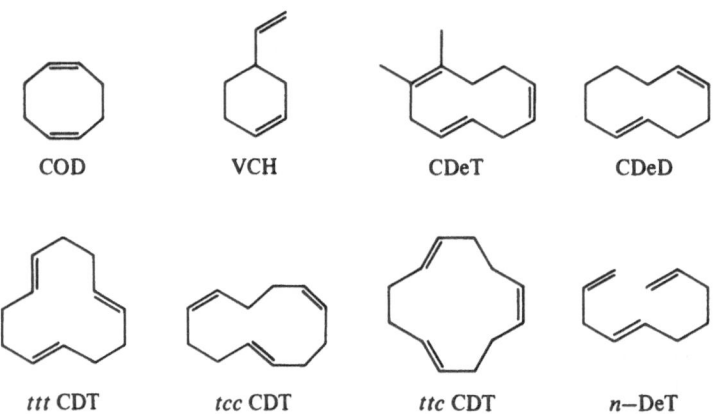

COD	VCH	CDeT	CDeD

| *ttt* CDT | *tcc* CDT | *ttc* CDT | *n*—DeT |

Scheme 1.–1. Products of metal-catalyzed ring syntheses

The objective of our group was to find out the scope and to get a further insight into the control of these catalytic syntheses. At the beginning, we intensively investigated further reactions of the catalytically formed rings[3–5] and the formation of mono- and di-methylsubstituted derivatives of the above mentioned rings[6]. New catalytic, carbocyclic ring syntheses as well as hetero-ring syntheses could be developed on the basis of Ni and other metals[7, 10] (Scheme 1-2). Also new open-chain oligomers could be formed catalytically[11–15] (Scheme 1-3).

Generally speaking, we had to face two fundamental problems of elucidating the scope and the control of the above mentioned reactions (Table 1.1).

The influence of properties and concentrations is manifold, especially including:
1) *Local structural and electronic situations* and their consequences for the total structure and reactivity in the light of the metala-logy principle[16–18]. *The control of the elementary steps* by the numbers and the properties of all participating species [f ( )] is of central interest (see Chap. 2).
2) The actual steady-state situation of the catalytic system:
   a) *Association-phenomena:* In order to find out *the associations of all participating species* in the catalytic process (better in all product-determining elementary steps of the cycles), caused by their properties [symbol f ( )] and concentrations (symbol f [ ]), we have applied and modified the "Method of Inverse Titration" to organometallic chemistry (Chap. 3)[19, 20] (for a review of titration procedures see Ref.[21]).

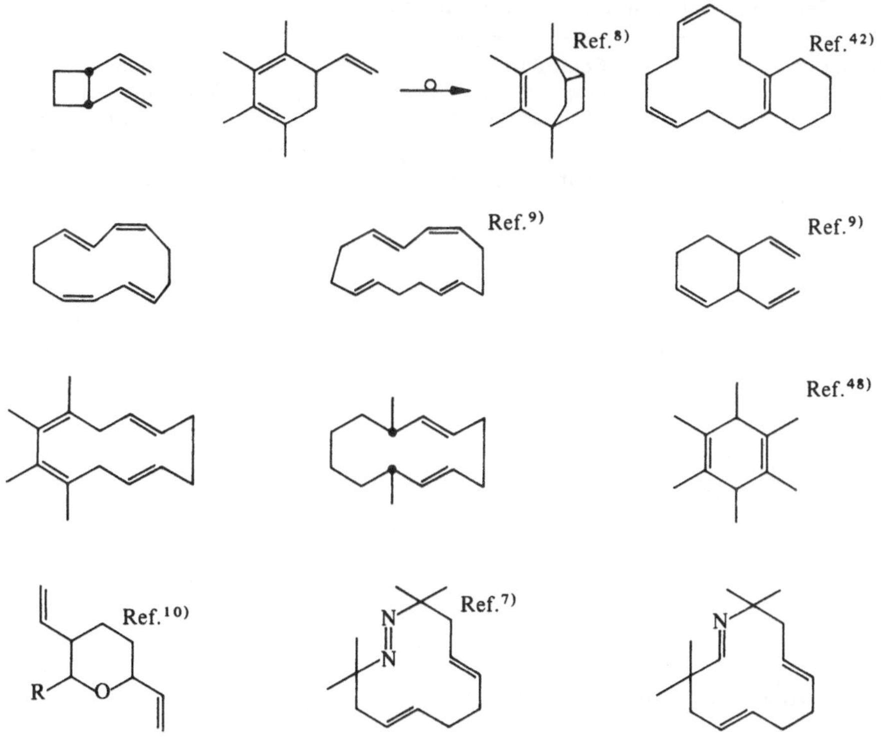

Ref.[8]) Ref.[42])

Ref.[9]) Ref.[9])

Ref.[48])

Ref.[10]) Ref.[7])

Scheme 1.–2. Further metal-catalyzed ring-syntheses

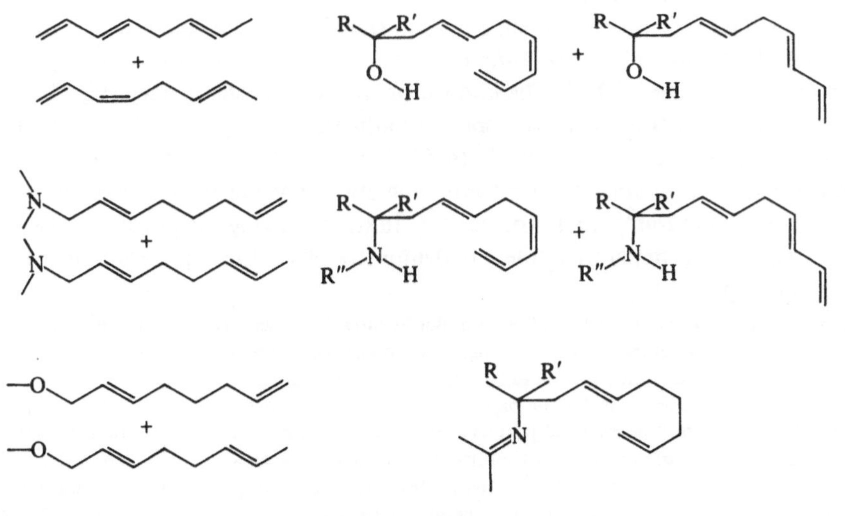

Scheme 1.–3. Catalytically formed acyclic products

47

Table 1.1. Controlling factors of metal-catalyzed processes

Controlling compounds	Control by properties of all species	Control by concentrations of all species
	symbol f ( )	symbol f [ ]
substrate	f (S)	f [S]
co-substrate	f (S')	f [S']
metal	f (M)	f [M]
ligand	f (L)	f [L]
co-ligand	f (L')	f [L']

b) *The distribution of the metal* to differently coupled cycles and "inactive" complexes and within the cycles the distribution of the metal to educt-(substrate-), intermediate- and product-complexes (Chap. 3) is important.

2 The Metala-Logy Principle

2.1 Introduction and Fundamentals

To understand the catalytic system under investigation as a whole, we tried to transfer experimental *methods* and the corresponding *formalism* from inorganic chemistry (see Chap. 3) and enzyme chemistry [22] to metal catalysis. To get an insight into the control of elementary steps in the catalytic cycles, we developed as an experimental tool the metala-logy principle[23], which allows to transfer to or to find out in organometallic chemistry some well known *rules* of theoretical chemistry applied in organic chemistry (see also Scheme 2.1-1). The knowledge of catalytic elementary steps is concentrated with respect to syntactic aspects (mathematical formalisms), e.g. in the so-called three-center rules of K. Fukui[24] (see Scheme 8 in Ref.[22]). The semantic aspects are especially shown by demonstrating the importance of donor-acceptor qualities, e.g. for structures[25] and processes[26]. In the following the pragmatic (experimental) aspects for preparative rules are demonstrated by the application of the metala-logy principle.

To introduce our model we first of all would like to show by general schemes the importance of "molecular architecture" of organic chemistry in experiment and theory (for theory see e.g. Refs.[27-29]). Two basic models of an experimentally working organic chemist are given in Scheme 2.1-2. In a model, we consider the organic moiety as a given basis (in Scheme 2.1-2 e.g. ethylene and benzene) modified by local perturbations *in* and *at* the system. A local perturbation e.g. *in* a π-system, involves substitution of C-atoms by heteroatoms such as N, O etc. and even metals. A local perturbation *at* a π-system involves substitution of hydrogen by functional groups. The combination of π-systems (see model for intermolecular perturbations) is not restricted to direct coupling, but π-systems can also interact by means of different couplers as shown by the coupling

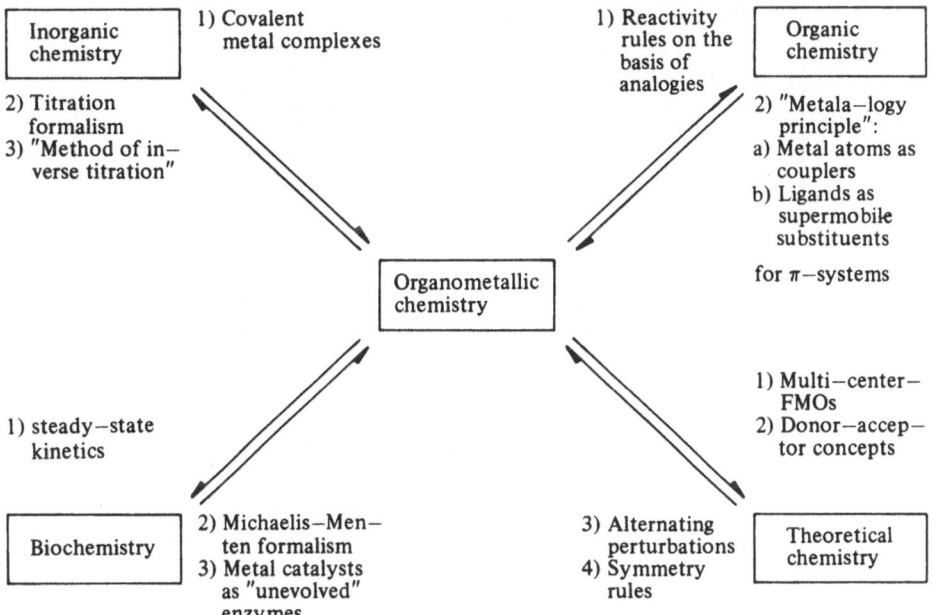

Scheme 2.1–1. Multi-methods and multi-rule models

of two ethylenes. The vinylogy- and phenylogy-principle [30] have shown their pragmatic importance for the preparative chemist in charge-controlled processes. In orbital-controlled reactions, these principles were extended by WH-rules. Other couplers have been discussed in organic systems (see e.g. Ref. [31]). By the metala-logy principle we would like to widen up these useful concepts to the unifier metal. Using the given analogies (see Ref. [32]), the following is aimed at:

1) better, perhaps the correct experiments and questions,
2) models which we can transfer to organometallic chemistry,
3) transfer of experience in organic chemistry to metal catalysis.

The importance and meaning of some of the models of Scheme 2.1-2 in a more theoretical aspect are demonstrated by some examples in Scheme 2.1-3.

When we transfer the theoretical aspects of these models to organometallic chemistry, we are aware of the fact that, e.g. when altering the number of π-electrons by two (6π-, 4π- or 2π-electrons) in the elementary steps of organometallic reactions, we have, step-by-step to await a change in the type of process (S or A) [33] (in analogy to Fig. 1 in Scheme 2.1-3). On the other hand, we also can learn that the energies of the FMO's change systematically, going from 6π- to 4π- to 2π-electron systems (from e.g. bis-π-allyl- to π-allyl-σ-allyl to bis-σ-allyl-metal-complexes).

From the organic chemistry of cyclic π-systems, we can transfer to metal-complexes e.g. the consequences of donor-acceptor perturbations by canceling the degeneracy of orbitals (compare Ref. [34]); analogies concerning Fig. 2 in Scheme 2.1-3 are found in Sect. 2.5). The generalized effects of local perturbations by classifying the substituents – in consequence of orbital energies and of changes of the coefficients in these orbitals [35] – may analogously be transferred to organometallic chemistry [36] (Fig. 3 in Scheme 2.1-3).

Fig. 1. Perturbation in and at ethylene or benzene

Perturbation	System ⟋★ Perturbation *at* the system	System ★ Perturbation *in* the system
Type	$CH_2=CH-CH_3 = -\ddot{X}^a$ $CH_2=CH-COOR = -Z$ $CH_2=CH-C_6H_5 = -C$	$CH_2=NH$ $CH_2=O$ $CH_2=S$ 
Position	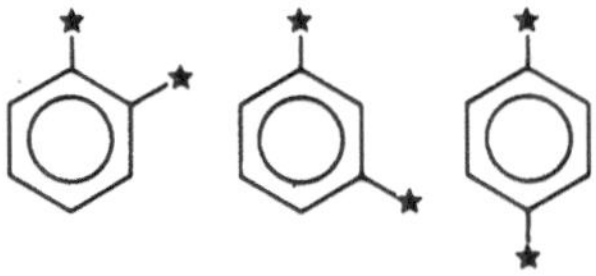	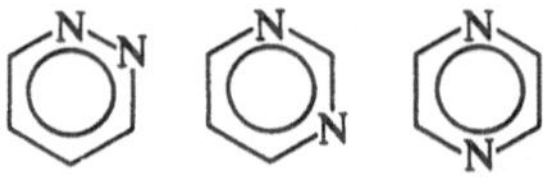
Number	$CH_2=C\begin{smallmatrix}CH_3\\H\end{smallmatrix}$ $CH_2=C\begin{smallmatrix}CH_3\\CH_3\end{smallmatrix}$	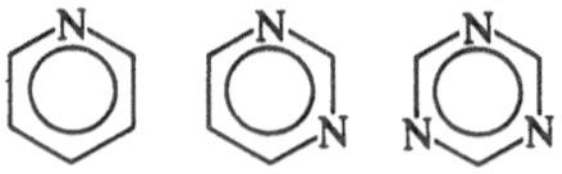

[a] see *Fig. 3* in *Scheme 2.1–3*

Fig. 2. Combination of π-fragments through couplers

| Fragment | Coupler | Fragment |

Type of unifying for ethylenes			
Coupler	Examples	Symbol	Trivial name
direct	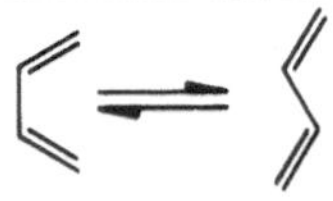	◄—u—►	Unifying principle
•CH$_2$ •CH$_2$		◄—u$_e$—►	Ethylogy principle
	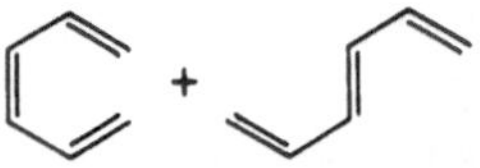 +	◄—u$_v$—►	Vinylogy principle
	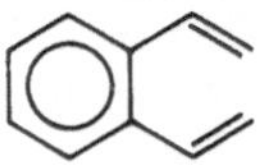	◄—u$_{ph}$—►	Phenylogy principle
 ◐M A[a] ◯M S[a]	"s–cis" X 	◄—u$_m$—►	Metala–logy principle

[a] A = antisymmetric; S = symmetric (see Ref.[33])

Scheme 2.1–2. Substitution *in* and *at* the system

Fig. 1. see Ref.[28b]) Model for open-chain π-systems   **Fig. 2.** Model for cyclic π-systems

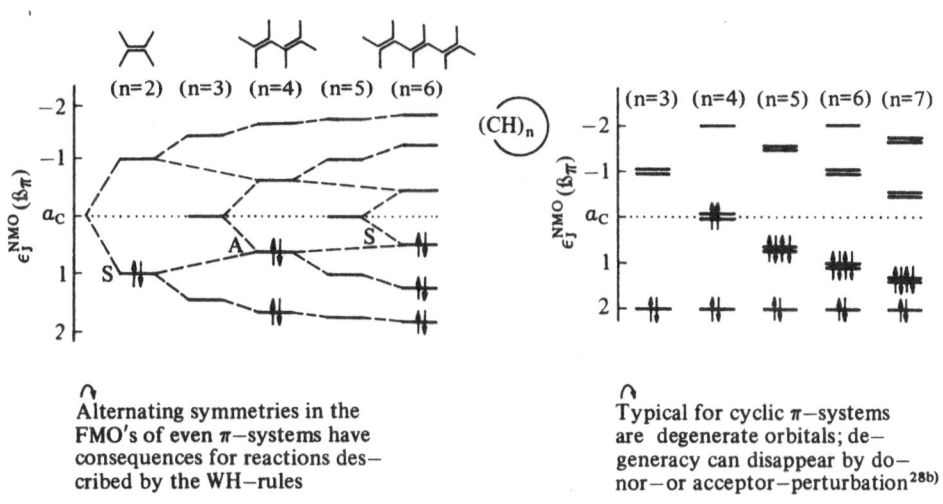

Alternating symmetries in the FMO's of even π–systems have consequences for reactions des–cribed by the WH–rules

Typical for cyclic π–systems are degenerate orbitals; de–generacy can disappear by do–nor–or acceptor–perturbation[28b]

Fig. 3. Generalized effects of local perturbations on FMO's

Shown for ethylene[35])

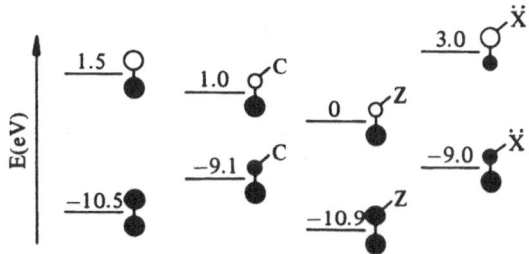

The opposite effects of $\ddot{X}$ (e.g. CH_3) and Z (e.g. COOR) or the C–type perturbation (e.g. C_6H_5) on the coefficient c in LUMO (but the same effect on c at the centers in HOMO) allow the experimental determination of HOMO– or LUMO–control by regio–isomeric distribution in orbital–controlled processes c^2 is a measure of the electron demand in LUMO and the electron density in HOMO)

Scheme 2.1–3.

Figure 1 in Scheme 2.1-4 shows that, instead of *dual thinking* (e.g. in HOMO/ LUMO- or LUMO/HOMO-control for structures — considering the structural con- sequences for the joining of molecular fragments — or processes), we also have to take into account the consequences of relative donor-(DO) or acceptor-perturbations (ACC) in the FMO's of π-systems (*double-dual thinking*).

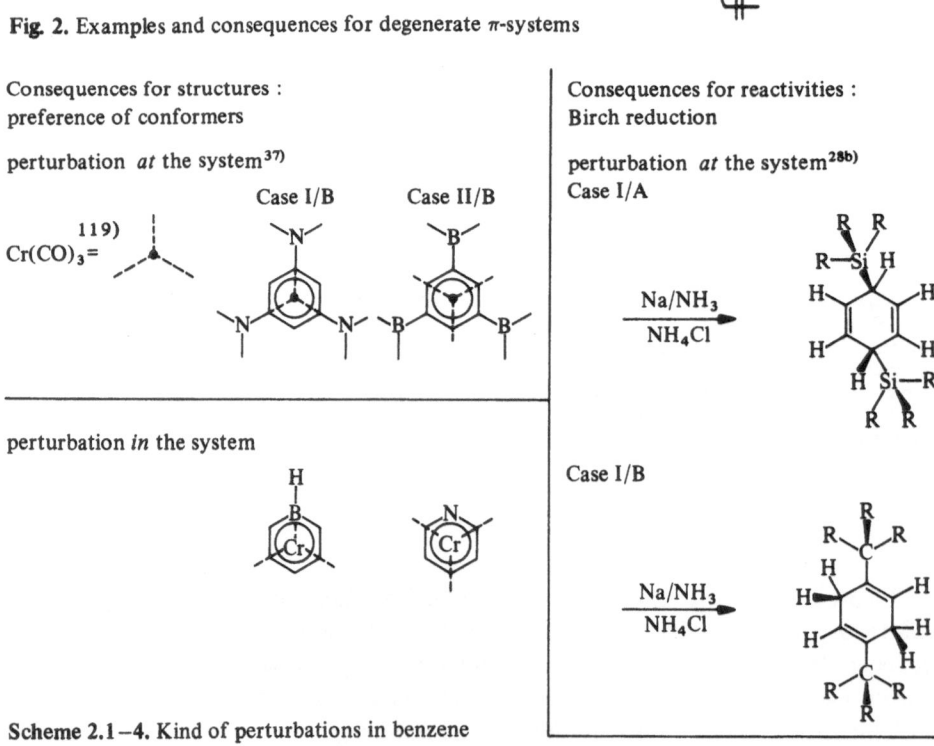

Fig. 1. Kind of perturbations in benzene

$B \triangleq$ benzene
$P \triangleq$ perturber

$\underline{HO_B^{DO} / HO_P^{ACC}}$ (Case I/A)

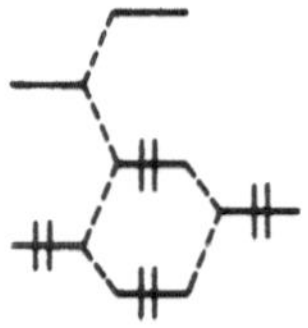

Case I
$\underline{LU_B/HO_P}$

$\underline{HO_B^{ACC} /HO_P^{DO}}$ (Case I/B)

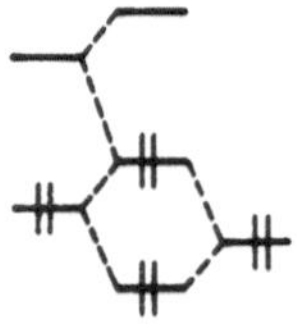

119)
$\underline{LU_B^{DO} /LU_P^{ACC}}$ (Case II/A)

Case II
$\underline{HO_B/LU_P}$

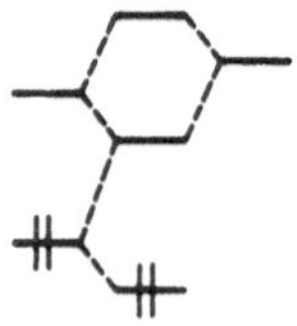

$\underline{LU_B^{ACC} /LU_P^{DO}}$ (Case II/B)

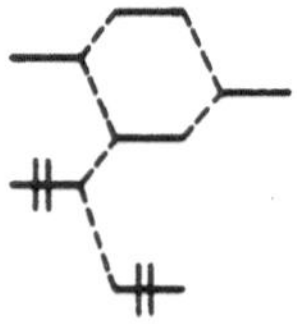

Fig. 2. Examples and consequences for degenerate π-systems

Consequences for structures :
preference of conformers

perturbation *at* the system[37]

Cr(CO)₃ = 119)

Case I/B   Case II/B

perturbation *in* the system

Consequences for reactivities :
Birch reduction

perturbation *at* the system[28b]
Case I/A

$\xrightarrow[NH_4Cl]{Na/NH_3}$

Case I/B

$\xrightarrow[NH_4Cl]{Na/NH_3}$

Scheme 2.1–4. Kind of perturbations in benzene

The HO-energy of a π-system can be changed, e.g. by occupied orbitals of hetero-atoms which are relative donors or acceptors for the HO's. This can also be true of interactions of the LUMO's of the π-system with vacant AO's of heteroatoms. The consequences e.g. for the preference of certain conformers relative to the type and position of perturbations in tricarbonyl-chromium benzene complexes [37] (organo-metallic example) are described in Figure 2 of Scheme 2.1-4 together with the conse-quences for the reactivities in benzene derivatives (example of organic chemistry) due to relative donor- or acceptor-perturbations [28b] (see also Scheme 2.1-2 Fig. 2).

As is evident by the given examples in Scheme 2.2-2 we would like to restrict our-selves to systems in which MO- and especially FMO-control and no charge and "ste-ric" control plays a significant role (but see Sect. 3.5).

As a rough, pragmatic idea we assume, that for even carbon π-systems, hetero-atomic substituents introduce HO- or LU-interactions (dual systems) and additional DO/ACC-perturbations *in* or *at* the system (double-dual system). Comparing schemat-ically the HO's and LU's of carbon π-systems with even number of π-electrons and the relative energies of the highest occupied orbital at the central heteroatoms of a functional group *at* the system, we can imagine, how to realize relative DO/ACC per-turbations. The interactions of the FMO's of acids and bases with the FMO's of the carbon π-system (called intrapacket interactions[38]) lead, due to their relative DO- and ACC-properties, to consequences for the orientation in space[120].

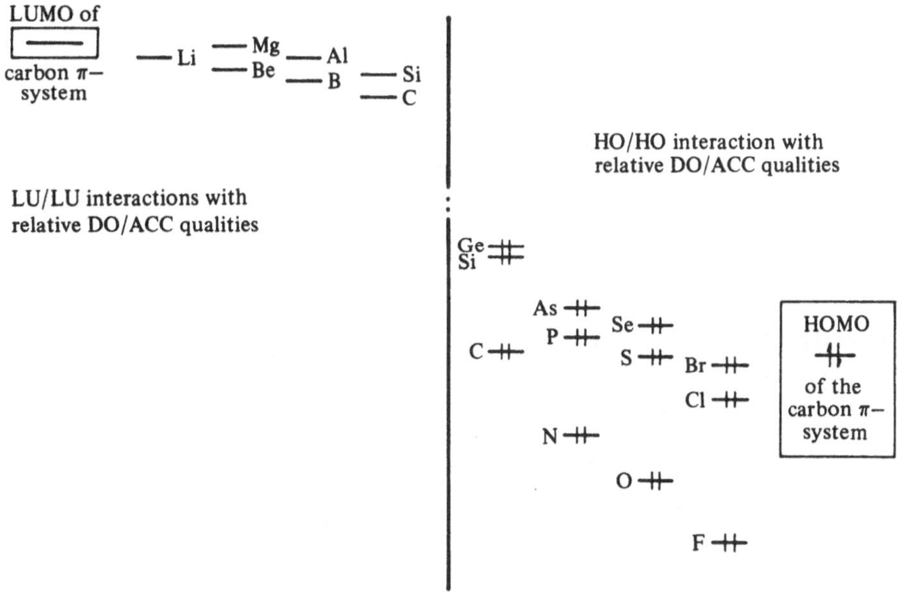

LUMO of
carbon π-system

—Li —Mg —Al —Si
     —Be —B   —C

LU/LU interactions with relative DO/ACC qualities

HO/HO interaction with relative DO/ACC qualities

Ge
Si

As Se
P   S
C   Br
Cl
N
O
F

HOMO
of the carbon π-system

2.2 The Empirical Model

Our basic empirical concept is that chemical systems coupled via a metal atom be-have like fragments directly joined with one another by σ-bonds. In Scheme 2.2-1 the formal procedure for the linkage of organic moieties is compared with that re-

P. Heimbach and H. Schenkluhn

quired by the metala-logy principle. Connecting two ethylenes to butadiene is not a real reaction but only model-thinking. The coupling of two ethylene molecules to a "metala-butadiene" is a model, too, but can easily be realized by association of two

Fig. 1. Procedure in unifying principle: a model

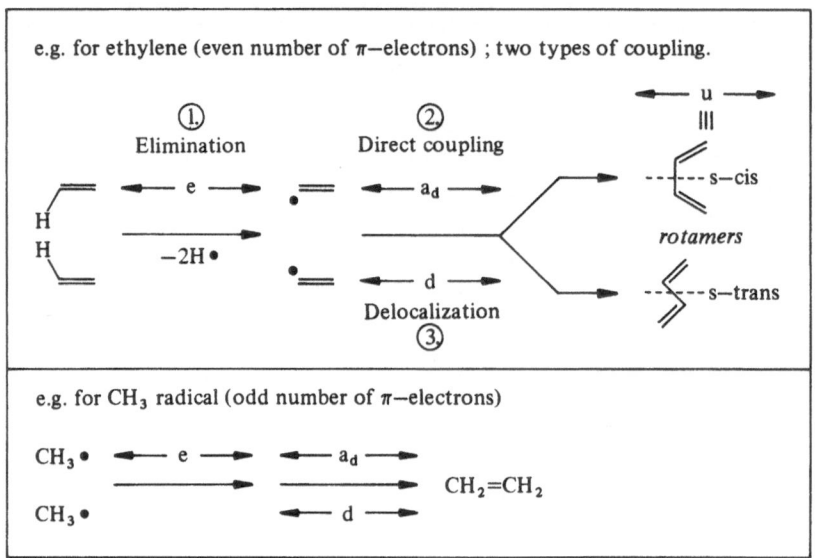

Fig. 2. Procedure in metala−logy principle: a reaction

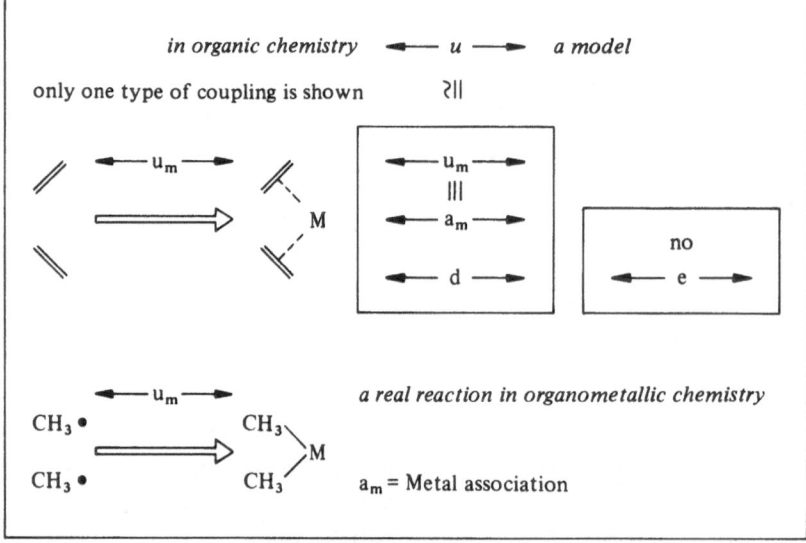

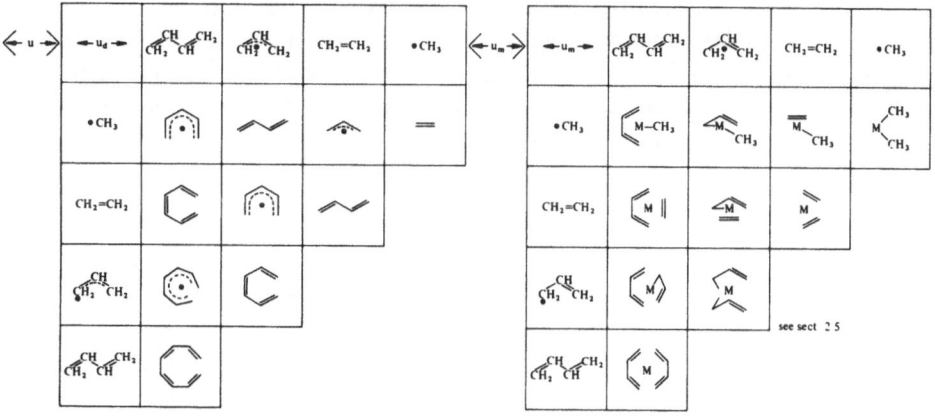

Scheme 2.2—1. Procedure in unifying principle: A model

ethylenes on a metal. However, there is one big difference when comparing e.g. two allyl groups coupled via a C—C-σ bond or via a metal. When linked via a C—C-σ bond with local C_2 symmetry, the two ends of the trans π-system (two-dimensional π-sys-

tem) do not interact so that e.g. α, ω-ring closure does not occur. This may, however, be possible in the metala-analogous bis-π-allyl-metal complexes. We would especially note that these complexes contain three-dimensional π-systems (see also Ref.[31]).

Also in organic chemistry there are examples illustrating that planarity is not a prerequisite for intensive π-interactions (see e.g. non-planar π-interactions via a four-membered ring[31]).

At the bottom of Scheme 2.2-1 a general comparison of directy joined π-systems and metal-linked π-systems is given. (For a general discussion of the number of unifying procedures according to the metala-logy principle see Sect. 2.5). An interesting consequence of this comparison is that e.g. allyl-alkyl-metal complexes may react similarly and especially by a similar control like bis-olefin complexes both by forming C—C bonds (metala-ring-closure reactions):

55

To find out applications and rules of the metala-logy principle, we systematically investigated the influence of HOMO/LUMO- or LUMO/HOMO-perturbations (dual model) as well as relative donor/acceptor- or acceptor/donor-interactions (double-dual model) in metal-complexes [39−41], in their model-reactions [39, 40] and in metal catalysis: The strategy we followed is depicted in Scheme 2.2-2. For our purposes, we separated the studied catalytic system into two π-systems (substrates S: Case I),

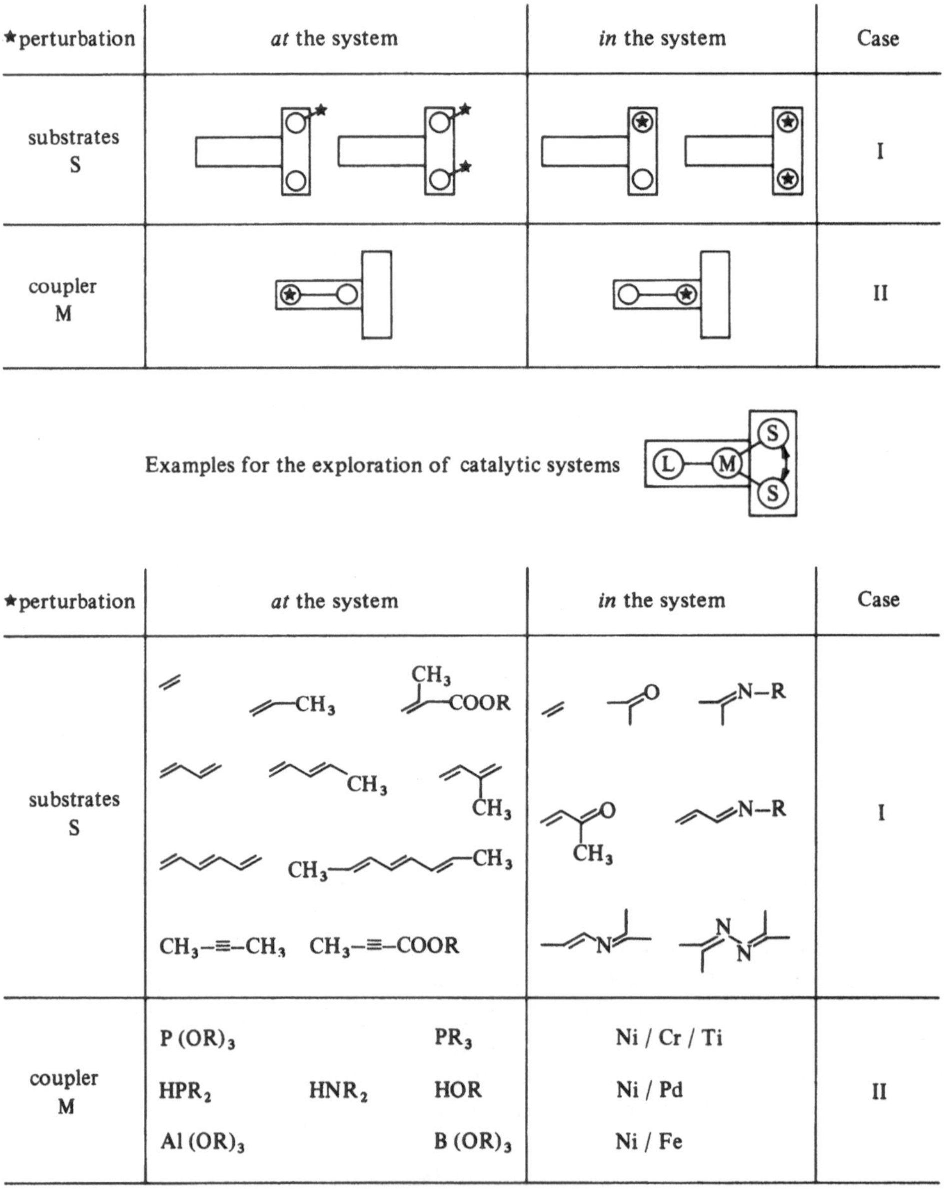

★perturbation	*at* the system	*in* the system	Case
substrates S			I
coupler M			II

Examples for the exploration of catalytic systems

★perturbation	*at* the system			*in* the system		Case
substrates S						I
coupler M	$P(OR)_3$		PR_3	Ni / Cr / Ti		II
	HPR_2	HNR_2	HOR	Ni / Pd		
	$Al(OR)_3$		$B(OR)_3$	Ni / Fe		

Scheme 2.2−2. Metala-logy principle: experimental strategy

which are joined by the metal, and the coupling metal M (Case II), which is modified by further ligands L — here simplified to only one ligand (for more detailed information on ligand associations see Chap. 3).

Case I. First of all, we systematically introduced perturbation *at* coupled π-systems (e.g. ethylene, butadiene, hexatriene and combinations of them) by substituting hydrogen for functional groups (classified by K. N. Houk as C-, Z- and X-substituents[35]), especially by methyl[6], phenyl and carboxylic ester groups. We have so far introduced perturbations in the π-system by replacing carbon atoms by heteroatoms (N and O) in the π-system skeleton.

Case II. By keeping the reacting substrates S constant, we varied systematically the coupler M (variation *in* the coupler), and the modifying ligands L (variation *at* the coupler). The influence of the metal M on structure and reactivity we tried to find out by comparing pairwise Ni- with Pd-, Ni- with Fe- or Ni- with Cr- and Ti-catalysts and so on. Our goal is to work out a schematic rationale for transition metals similarly as we try to analyze the control of substituents at the carbon π-system and the co-ordination of the ligands as DO/ACC bases or DO/ACC acids as shown on page 53. In Sect. 2.4 we will discuss as an example the influence of some Lewis bases, especially phosphorus containing ligands on catalytic processes. For a better understanding of catalytic reactions and metalorganic elementary steps and in order to find out catalytic systems, the following series of dual investigations are of great importance:
1. perturbation *in* and *at* the reacting carbon π-system 2. left and right side of PSE
3. first and second rows of the given elements.

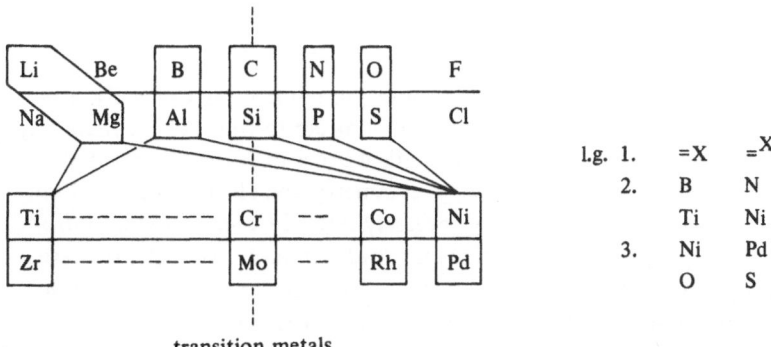

transition metals

Some of these combinations will be discussed in Sect. 2.3. (see[120]).

2.3 The Hierarchy of Problems in the Metala-Logy Principle and Experimental Evidence for the Existence of Pragmatic Rules

Scheme 2.3-1 demonstrates the hierarchy of selected questions to elucidate the controlling factors in the elementary steps of catalysis. With this scheme we would like to point to the fact, that it is necessary to distinguish various possibilities of property

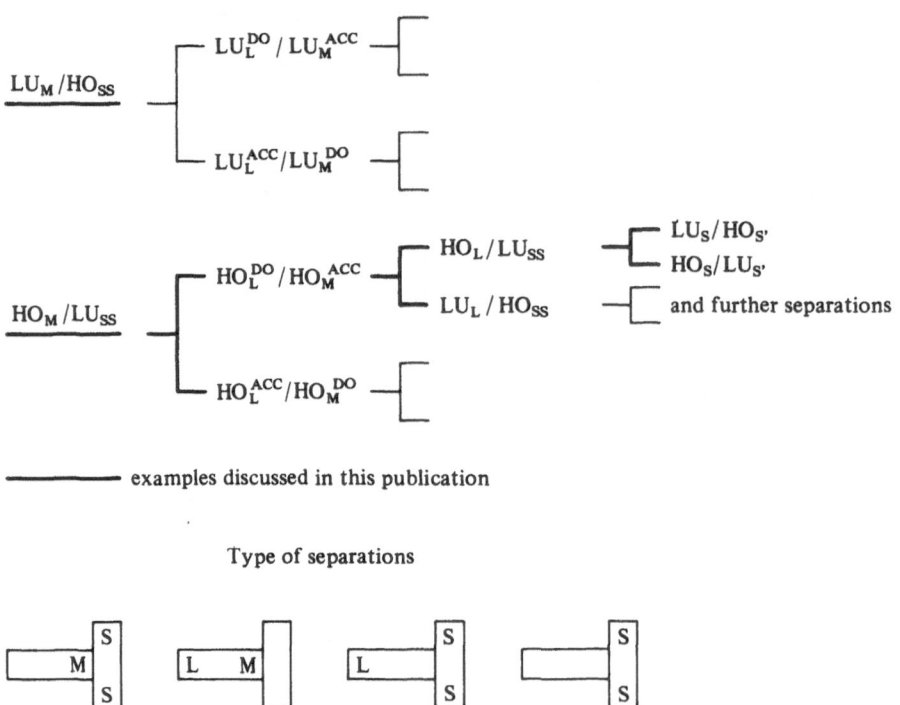

LU_M/HO_{SS}

LU_L^{DO}/LU_M^{ACC}

LU_L^{ACC}/LU_M^{DO}

HO_M/LU_{SS}

HO_L^{DO}/HO_M^{ACC}

HO_L/LU_{SS}

LU_L/HO_{SS}

$LU_S/HO_{S'}$
$HO_S/LU_{S'}$

and further separations

HO_L^{ACC}/HO_M^{DO}

————— examples discussed in this publication

Type of separations

Scheme 2.3—1. Hierarchy of problems in metal catalysis

specific control in catalytic systems (for concentration control see Chap. 3). The high complexity in catalysis has to be "digitalized" in order to find out the hierarchy of these controlling factors. Changing the hierarchical order of controlling factors, we expect alterations with respect to topology, symmetry and electron distribution in structures and processes and evidence for inversely related trends raised by introducing the same set of perturbations. All this may lead to alternative regio- or stereo-isomers within the products or also to characteristically changing preference within the combination of the competing elementary reactions. What individual controlling factor really determines the single decisions in the catalytic process has to be elucidated by independent investigations of model-complexes, their reactions and their individual couplings in the whole system.

In Tables 2.3-1 and 2.3-2 the products formed by Ti- and Ni-catalysts are compared with those formed by Cr catalysts. Ni- and Ti-catalysts show an opposite behavior in the *ttt/ttc*-isomer distribution of 1,5,9-cyclododecatrienes and in 2 : 1- vs. 1 : 2-adduct formation in the co-oligomerizations. The Cr catalyst behaves like a "mixture" of the Ni- and Ti-catalyst.

Cr-catalyst can perhaps "pair" the electrons of the half-filled d orbitals in two ways, leading to comparable control like in Ni- or Ti-catalysts. It can be assumed, that HO_{SS}/LU_M(Ti)- or HO_M/LU_{SS}-interactions (Ni) may lead to the observed consequences.

Table 2.3–1. Cyclotrimerization of butadiene by means of Ni-, Cr- or Ti-catalysts[5)]

Substrate	Products	Catalyst Ni	Cr	Ti
		95%[a]	60%	5%
		5%	40%	95%

[a] We added pyridine in a ratio pyridine:nickel = 1:10 to avoid tcc- CDT formation

Table 2.3–2. Co-oligomerization of a 2:1 mixture of butadiene and 2-butyne at 40 °C by Ni-, Cr- and Ti-catalysts (the product yields (%) are based on alkyne)

Substrate	Products	Catalyst Ni··L	Cr	Ti
		> 90%	30%	—
		—	5%	45%
		—	10%	30%

In Scheme 2.3-2 a selected number of reactions catalyzed by Ni- and, alternatively, Pd-complexes are compared with respect to different local symmetry of the dotted subsystem (local σ- or C_2-structure or process in Fig. 1 of Scheme 2.3-2) or with respect to the influence of alkyl groups on a fragment (subsystem) of the substrates resulting in different structures of the products (Fig. 2 of Scheme 2.3-2).

P. Heimbach and H. Schenkluhn

Fig. 1. Product with local σ- or C$_2$-symmetry

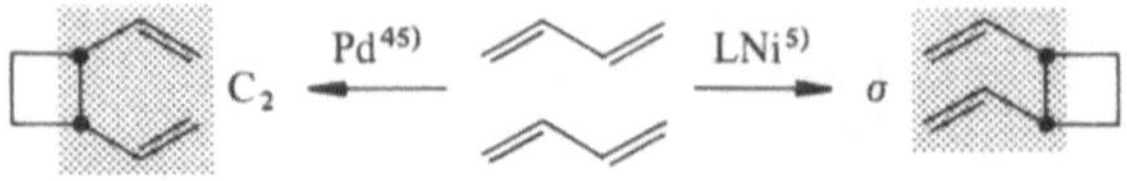

C$_2$ ← Pd[45] ← + → LNi[5] → σ

Fig. 2. Opposite influence of an inductive effect by alkyl groups

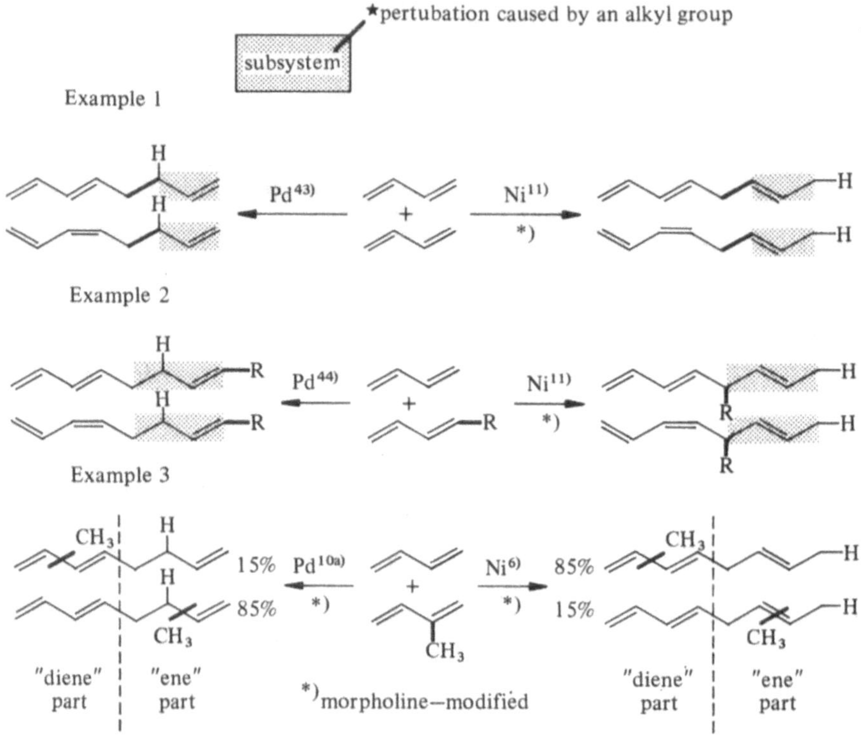

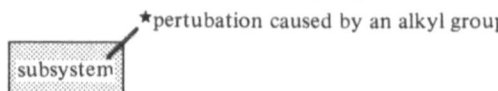

*pertubation caused by an alkyl group

subsystem

Example 1

H, H — Pd[43] ← + → Ni[11] *) → —H, —H

Example 2

H, H —R — Pd[44] ← + —R → Ni[11] *) → —H, —H, R, R

Example 3

CH$_3$, H, H — 15% Pd[10a] 85% *) ← + CH$_3$ → Ni[6] *) 85% 15% → CH$_3$ —H, CH$_3$ —H

"diene" part | "ene" part   *)morpholine—modified   "diene" part | "ene" part

Fig. 3. Different combinations of elementary steps

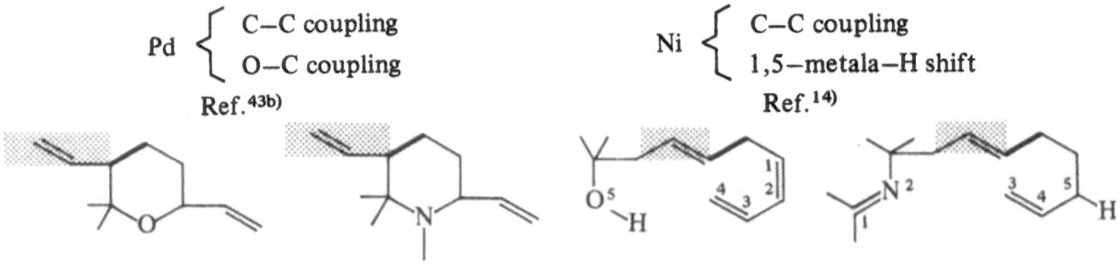

Pd { C—C coupling / O—C coupling   Ref.[43b]

Ni { C—C coupling / 1,5—metala—H shift   Ref.[14]

O^5—H   4 3 2 1   N^2 1 / 3 4 5 —H

Fig. 4. Established an assumed structures of Ni- and Pt-complexes

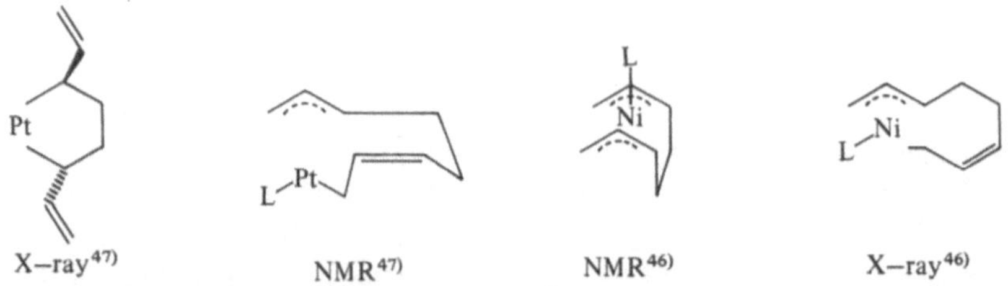

Pt   L—Pt   L—Ni   L—Ni

X—ray[47]   NMR[47]   NMR[46]   X—ray[46]

Scheme 2.3—2. Comparison of Ni- and Pd-catalysts

Figure 3 in Scheme 2.3-2 illustrates that Ni- or Pd-complexes prefer a different combination of elementary steps. Here, it is evident that Ni favors 2 : 1 co- oligomerization of butadiene with aldehyde or of a Schiff base with butadiene involving C—C-bond formation coupled with metalalogous 1,5-hydrogen transfer. On the other hand, Pd favors O—C- or N—C- and C—C-bond formation. These processes seem to occur more frequently, as demonstrated by other catalytic processes[41] and model reactions[41, 43]. Figure 4 in Scheme 2.3-2 reveals that with comparable Ni-[46] and Pt-intermediary complexes, differences in local symmetries may result. In addition, Ni promotes interactions with a primary carbon atom of the σ-allyl group whereas Pt interacts with the substituted carbon atom of this group.

Coupling of two π-systems by a metal atom raises further problems in the hierarchy shown in Scheme 2.3-1 namely whether the metal is e.g. a HO_M^{DO}- or a HO_M^{ACC}-coupler of the carbon π-systems and what changes experimentally result due to this alteration. These problems may perhaps be elucidated by comparing the above mentioned Ni- and Pd-catalyzed processes. The preliminary rule that e.g. the local symmetry in structure is different and that the same inductive effects (here of the alkyl groups) lead to opposite couplings is especially supported by model complexes shown in Fig. 4 of Scheme 2.3-2. For the given catalytic examples, other effects (e.g. changing associations, see Sect. 3.3) may lead to similar consequences, but this argument can be excluded for the well-defined complexes.

Another type of differentiation involves a cooperative change in the reacting substrates. This plays an important role in phosphine and phosphite Ni-catalysts

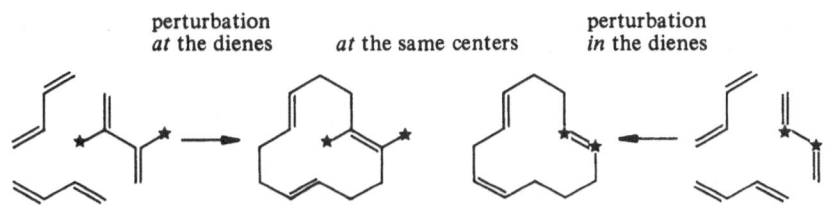

Isomer distribution in 2:1 co—oligomers (two 1,3—butadienes and one 1,3—diene)

2 : 1 derivatives	1,3–dienes				
	H ... H	CH₃ ... CH₃	(cyclohexene diene)	(N compound)	(N–N compound)
ttt CDT	83%	55%	– – –	99%	100%
tcc CDT	11%	9%	– – –	– – –	– – –
ctt CDT	6%	36%	80%	1%	– – –
ccc CDT	– – –	– – –	20%	– – –	– – –

$Ni^{(o)}$: butadiene : diene = 1 : 200 : 100; reaction temperature : 40°C

Scheme 2.3–3. Influence of perturbations on Cyclo-cooligomerization compared with 1,4-diazadiene-modified Fe-catalysts [48]. The next step forward in the hierarchy of problems (Scheme 2.3-1), namely the influence of relative DO/ACC perturbations in substrates and ligands, is at best a twofold one, even when the metal and its oxidation state remain unchanged. In one series of experiments, the properties of the substrates were kept invariable while the properties of the ligands in constant associations (see e.g. Sect. 2.4 and type ⓓ analysis in Sect. 3.4) were

Fig. 1. FMO's of perturbed ethylenes[49)]

π^*CC —— 0.2533 π^*CC —— 0.242

π^*CN —— 0.1551

0.5690 a.u.   0.6135 a.u.   0.5921 a.u.

πCC —— -0.3157 πCC —— -0.3709

πCN —— -0.4370

$$CH_2C{=}\overset{\overset{\textstyle CH_3}{|}}{C}H$$ $$H_2C{=}CH_2$$ $$H_2C{=}NH$$

Fig. 2. Balanced perturbations *in* and *at* 1,3-dienes *at* different centers

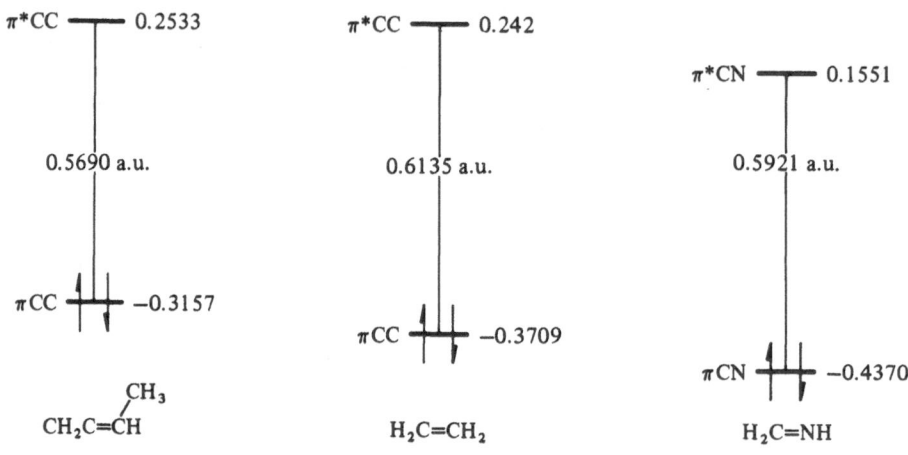

2x 1,5–metala–H–shift
2x C–C coupling ◄──►

more or less balanced situation in a catalytic process:

The ratio of diaza– to tetraaza–CDT depends on the butadiene to azine ratio and on DO/ACC qualities of the ligand

Fig. 3. Balanced and unbalanced situations in different steps of the catalytic cycles[13,41]

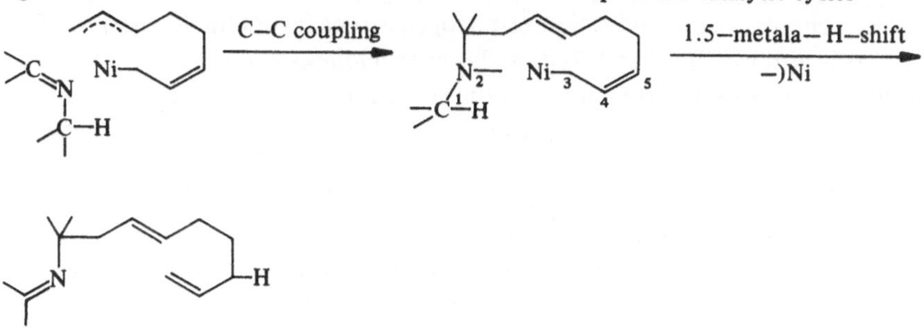

Fig. 4. Influence of DO/ACC perturbation in the cosubstrate on the ratio of synthon coupling[5]

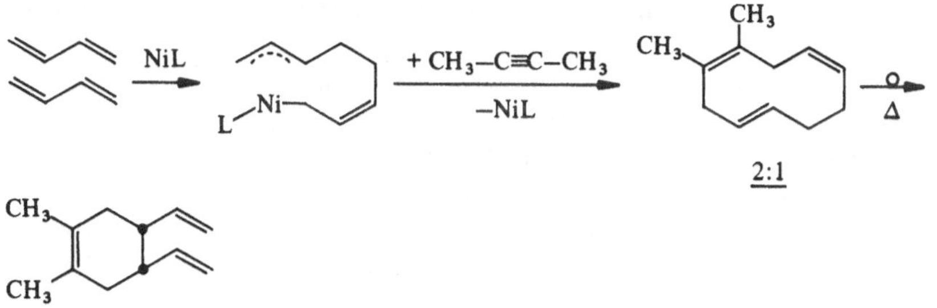

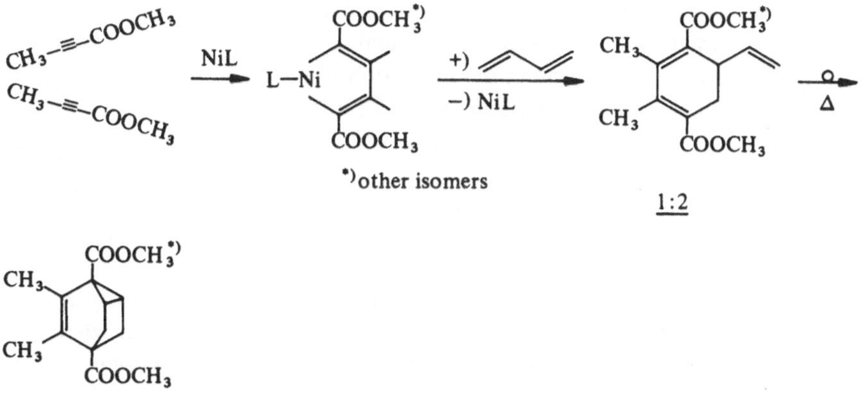

*) other isomers

Scheme 2.3–4. DO/ACC-Perturbations in substrates

changed. In a second series, the properties and associations of the ligands (see Sect. 3.5) were kept constant and the properties of the substrates (both by perturbations *in* and *at* the systems) varied.

In Scheme 2.3-3 some examples of the influence of DO/ACC-perturbations *in* and *at* the catalytically reacting π-systems are given. It reveals the influence of CH_3- or other alkyl-groups *at* the π-system and that of the nitrogen atoms *in* the π-system for the Ni-catalyzed 2 : 1 co-cyclotrimerization of butadiene with these substrates.

Within the two groups of isomers formed, the alkyl perturbation favors the formation of ttc isomers whereas the introduction of nitrogen atoms favors the formation of ttt isomers. It is noteworthy that 1,2- dimethylenecyclohexane reacts with two butadiene molecules to form the ttc- and even the substituted ccc isomer (observed for the first time in a catalytic reaction — for unresolved problems see Sect. 2.5) rather than the tcc isomer.

In the co-oligomerization of π-systems, DO/ACC-perturbations of substrates (Houks [35] X̌- and Z-substituents) play a significant role (for the perturbation induced by Houks C-type substituents see Ref. [84]). The following two aspects will be discussed:

1) DO/ACC-perturbations within the same π-system can be balanced so that the molecule with the "balanced perturbation" behaves like the original one.
2) DO- or ACC-perturbations in one of two π-systems can change the molecular ratio in the co-oligomerization of these π-systems.

Following a calculation [49], a perturbation caused by the nitrogen atom *in* the ethylene molecule is stronger and opposite than that of a CH_3 group *at* the ethylene (Fig. 1, Scheme 2.3-4). Therefore, we first of all introduced nitrogen atoms in respectively the 2- and 2,3-positions of 1,3-dienes, because here the coefficients in the FMO's are low. We balanced these perturbations by introducing *at* the system two or four alkyl groups in the 1,4-positions where the coefficients in the FMO's are high. The strategy of "balanced perturbation" in this form is valid only for the substrate itself but not for all steps within the catalytic cycle. Therefore, we cannot expect an unchanged behavior for all steps of the hierarchically ordered controlling factors. Thus, in the dimerization of the modified substrates, metala-analogous 1,5-sigmatropic shifts play a significant role resulting in the formation of new heteroring systems [41]. But we succeeded in co-trimerizing butadiene and 2-mono- or 2,3-diaza-dienes to twelve-membered heteroring systems [7, 41] (Fig. 2 in Scheme 2,3-4). Alkylated C=N double bonds react like ethylene with butadiene to 1 : 2 adducts (Fig. 3, Scheme 2.3-4). Hydrogen-transfer reactions, as mentioned before, are favored in this case over N—C-bond-forming reactions, so that open-chain instead of heterocyclic products [13, 14] are formed.

Figure 4 in Scheme 2.3-4 demonstrates that when using a triphenylphosphane-modified Ni-catalyst, butadiene reacts with 2-butyne to form a 2 : 1-adduct whereas with methyl 2-butynoate, a 1 : 2 co-oligomer is obtained. Butadiene and phenylacetylene also form 1 : 2 products [8]. As we may have shown, a change from X̌- to C- or Z-type substituents in the co-substrates alters the ratio from 2 : 1 to 1 : 2 in a synthon coupling reaction.

In one example, we finally describe how DO/ACC perturbations *in* a model system will influence its structure and reactivity.

In organic chemistry, replacement of oxygen by sulfur leads to "Umpolung" phenomena [50]. We have selected this O- and S-perturbation because there is a relative wide gap for the HO_π system so that HO^{DO}- vs. HO^{ACC}-perturbations *in* the π-system can easily be realized (see schematic representation on page 53). (Ni part of the double-dual system shown below[39, 41]).

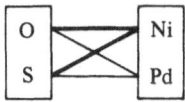

    Some selected results are shown in Scheme 2.3-5; these can be compared with differentiations in catalytic processes[39].

1) Change in local symmetry:
   Going from DO- to ACC-perturbation *in* the system, the structure of the dimeric species changes from the "chair"-(OCH_3) to the "boat"-form (SCH_3) (X-ray investigations[51]).
2) Further structural differentiation (in the allyl group; *anti/meso* vs. syn position):
   Both structural isomers show in their ^{1}H-NMR spectra two further isomers (OCH_3 = 50:50 and SCH_3 = 60:40) (see also Sect. 2.5). In complexes containing OCH_3-groups these two isomers differ especially in the δ-values of the anti hydrogens and in SCH_3-group containing complexes in the δ-values of the hydrogens in the syn- and meso-position of the allyl group.
3) Opposite results using the same set of changing inductive effects:
   At constant warming-up rates in well-defined solutions (0,05 M)[39] the thermal stability of OCH_3-group containing complexes increases whereas that of SCH_3-containing complexes decreases with rising inductive effect of alkylsubstituents in the allylgroups.
4) Differentiation in competing elementary reactions:
   The thermal decomposition of the OCH_3 complexes favors the pentene formation whereas that of SCH_3 complexes leads to a 1:1 ratio of pentenes to pentadienes.
5) Opposite symmetry control for the reaction pathways:
   In OCH_3 complexes, the trans isomer of 2-pentene while in SCH_3 complexes the cis isomer is favoured, starting from the same syn, syn-arrangement of the 1,3-dimethyl-allyl group.

2.4 Properties of Phosphorus-Containing Ligands as a Source of Information in Metal Catalysis and Model Complexes

Phosphorus-containing ligands show manifold influences on oxidation states and isomerisms in complexes and the extent of associations (see Chap. 3). The directing influence of varying properties in phosphines and phosphites was experimentally investigated for allyl-Ni-ligand complexes, e.g. with respect to:
1) the heat of formation of 1:1-NiL adducts[52],
2) the thermal stability of these 1:1 adducts[53],
3) the selectivity and reactivity of 1:1 complexes[53],
4) by ^{13}C-NMR data of the carbon atoms in the allyl group of 1:1 moieties[40] (for (2) and (3) see Scheme 3.5-5).

    On the other hand, the manifold influence of the varying properties of phosporus-containing ligands is evident, e.g. in catalytic
1) oligomerization and co-oligomerization of olefins and alkynes[4, 5], (see Schemes 3.5-2 und 3.5-4)
2) propene-dimerizations and -oligomerizations[5, 54],
3) oxo-processes to linear aldehydes[55],

4) hydrocyanation of butadiene [56],
5) butadiene polymerizations [57],
6) optical inductions [54, 58],
7) hydrogenations [59].

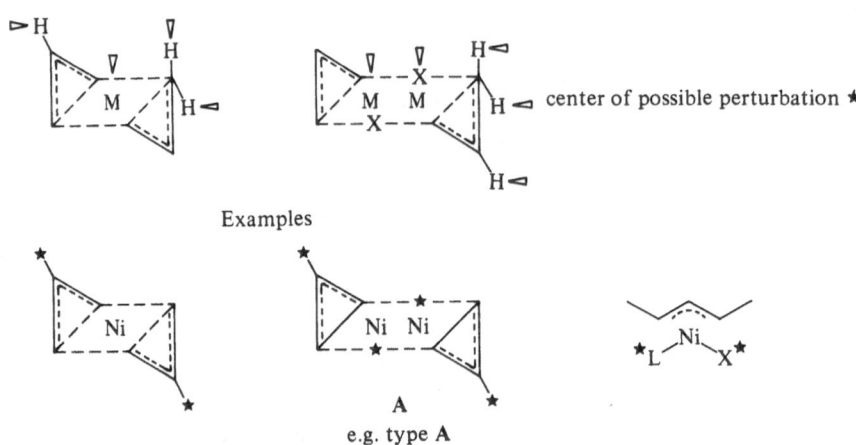

center of possible perturbation ★

Examples

A

e.g. type A

Structural change from "chair"—to "boat"—form

X—ray analysis of A

"Chair"   "Boat"

Further evidence for isomeric differentiations

NMR analysis

Similar δ values for syn—
and meso—protons but marked diffe—
rences for those of anti protons

Similar δ values for anti—
but clearly different ones between
meso—and syn—protons

syn δ = 1.99 (d)

anti δ = 1.34 (d)
and 1.27 (d)

OCH₃

syn δ=3.07(d)
2.96(d)

anti δ=2.05(d)

SCH₃

Scheme 2.3—5. DO- versus ACC-perturbations in model-π-systems[39]

Thermal stability of A in defined solutions

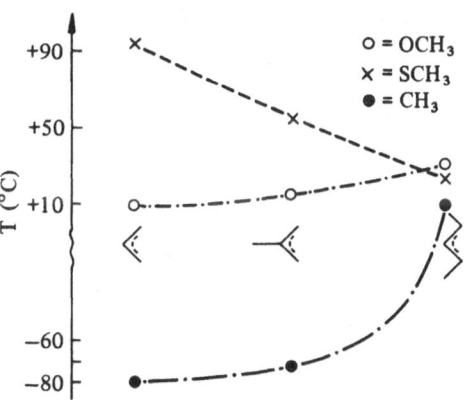

↖★ ⇨ ★↙ Ni /2 ⇩	Br	OCH₃	N(CH₃)₂
CH₃–⟨ Ni	(+53)*	+16	+84
H–⟨ Ni	+38	+10	+76
Ph–⟨ Ni	+11	+4	+66
Br–⟨ Ni	−14	−−	+38

*Chloro complex
Thermal stability of 2–substituted dimeric π–allyl–Ni–X complexes in 0.05 molar toluene solution (dec. temp. in °C).

Change of thermal stability with increasing methyl perturbation in dimeric π–allyl–Ni–X complexes in 0.05 molar toluene solution depending on type of X (dec. temp. in °C).

Metala–logous sigmatropic 1.5–shifts in A

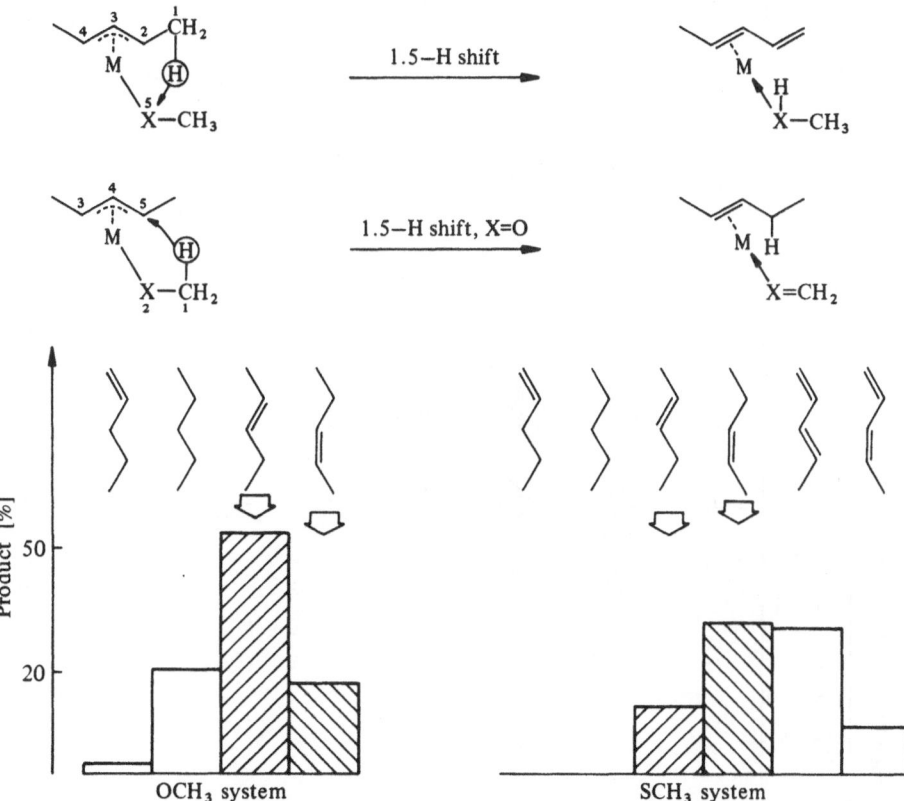

Scheme 2.3–5. (Continuation)

Table 2.4–1. Correlation of Tolman's electronic parameter χ with other parameters from transition-metal chemistry (II), organic chemistry (III) and phosphorus organic chemistry (IV)

	Parameter	Central atom	Def. value	Compound	Individual L's	n^a	Correl. factor	Ref.
I.	χ	Ni°	$\nu_{C\equiv O}$	$NiL(CO)_3$	70			60)
II.	χ^5	Mo°	$\nu_{C\equiv O}$	$MoL_2(CO)_4$	16	1	0.9943	64)
	σ	Mo°	$\nu_{C\equiv O}$	$MoL(CO)_5$	17	1	0.9738	65)
		Rh^I	$\nu_{C\equiv O}$	$RhL_2(CO)Cl$	24	1	0.9931	66)
		Ni°	$\nu_{N\equiv C}$	$NiL_2(tBuNC)_2$	8	1	0.9900	67)
		Co°	$\nu_{N=O}$	$CoL(NO)(CO)_2$	14	1	0.9600	68)
		Co°	$\nu_{C\equiv O}$	$CoL(NO)(CO)_2$	14	1	0.9915	68)
		Ni°	$\delta_{C=O}$ (^{13}C)	$NiL(CO)_3$	17	1	0.9868	69)

III.	σ_m	c	[aryl]—COOH	pK_a(rel)	16	1	0.9790	70)
	σ_I	c	*—CH_2COOH	pK_a(rel)	16	1 / 3	0.9666 / 0.9823	71)
	σ*	c	*—CH_2COOR	$lg\ k/k_0$	17	1 / 3	0.9809 / 0.9954	72)
IV.	σ	P	$R{-}P(={}O){-}OH$	pK_a	25	1	0.9428	73)

a $\quad \sigma_i = \sum_{j=0}^{n} a_j x_i^j$

Table 2.4–2. Correlation of Tolman's steric parameter with other steric parameters, using the isosteric principle[74, 75]

Parameter	Correlations	Definition value	Individual L's	n^a	Correl. factor	Ref.
Θ	R_3P vs. R_3P	cone angle	45			62)
E_s^S	R_2S vs. R_2PH	Pd^{II} (1,3-diene)OH_2^{2+} + $R_2S \longrightarrow$	6	1	0.9945	76)
E_s^N	R_3N vs. R_3P	$PhCOCH_2Br + NR_3 \longrightarrow$	16	1	0.9823	77)
E_s^C	R_3C vs. R_3P	$RCH_2COOH + CH_3OH \longrightarrow H + CH_3OH$	26	1	0.9662	78)

a $\quad E_s = \sum_{j=0}^{n} a_j \Theta_i^j$

In stoichiometric and catalytic processes information from the directing ligands is normally transferred to structure and reactivity of the associates. The advantage of catalytic reactions is that, cycle by cycle, this information can be accumulated. Therefore, it has always been of interest to find out relevant parameters for properties of phosphanes and phosphites by which the induced control can be transformed into a numeric form with predictable power.

Recently, *C. A. Tolman* successfully proposed two sets of parameters, the electronic Parameter χ [60] and the steric one Θ [61], for phosphorus-containing ligands. Their general importance is demonstrated by their correlations shown in Tables 2.4-1 and 2.4-2. The contribution of these two parameters to the control of the investigated elementary steps can be separated and their relative importance determined by a multilinear regression analysis [52, 63], leading to three-dimensional profiles analogous to Tolman's "steric and electronic box" [62] (see Sect. 3.3 and 3.4). In the interpretation of these profiles one should notice, that Tolman's parameters can linearly be correlated with the well-known Taft- and Hammett-parameters of organic chemistry [40] (Tables 2.4-1 and 2.4-2). From the shapes of the non-linear correlations of our experimental data with χ — when the electronic χ and steric Θ influence are separated by regression analysis — it may be concluded, that, besides donor/acceptor qualities of the P-ligand relative to the carbon-π-system (dual control), further double differen-

Fig. 1. Different types of mole‐cules of hexaphenyl—carbodi—phosphorane: X—ray evidence [79]

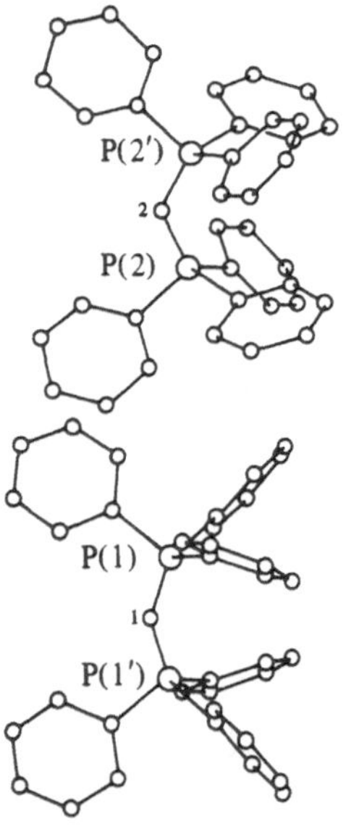

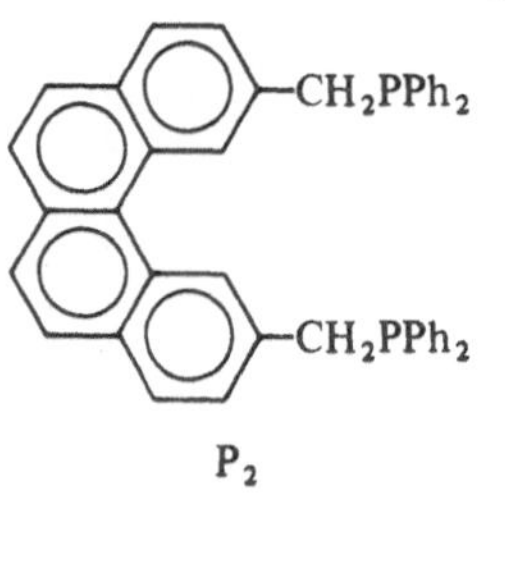

Fig. 2. Different types of molecular arrangements of the ligand P_2 in Ir^I— and Pt^{II}—complexes: X—ray evidence [80, 81]

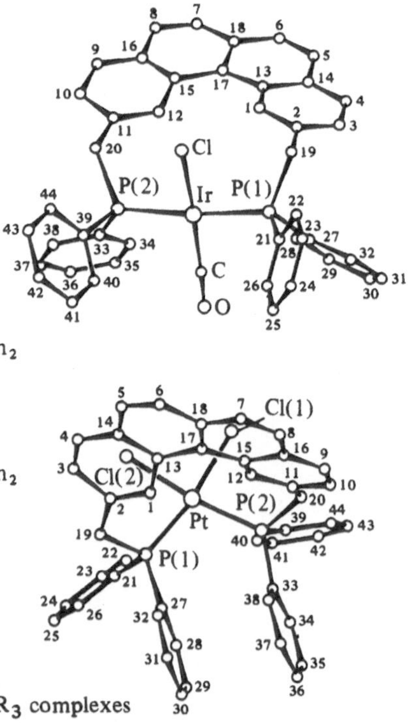

Scheme 2.4–1. Further evidence for differentiations in PR_3 complexes

tiation (double dual control) has to be taken into account involving information coming from the ligand itself (see local symmetries at L in Scheme 2.4-1).

In the following the importance of differentiation in local symmetry of phosphorus ligands will be demonstrated by an example out of phosphorus organic chemistry [79] and complex chemistry [80, 81] (Scheme 2.4-1). Typical of both systems is that C_3-symmetry reduction is twofold, leading to local different symmetry. The importance of this twofold differentiation for the orientation in space (e.g. optical induction) should be considered.

2.5 Some Aspects of the Metala-Logy Principle of Theoretical Interest

In this section we will briefly discuss some open problems. The metala-logy principle [85] leads to new chemical relationships of preparative importance. The concept of "isolobal fragments" (the metal with all ligands except the combined π-systems), developed by *R. Hoffmann* and coworkers [25, 26], will be extremely helpful from the theoretical standpoint. On the other hand this empirical principle raises some questions, the answering of which will acquire preparative relevance.

Question 1

Bis-allyl-metal complexes exist in three basic types of structural isomers: bis-π-allyl-, π, σ-bis-allyl- and bis-σ-allyl-moieties. Taking this into consideration and applying our principle results in the analogies shown in Scheme 2.5-1. These analogies are only complete if one con-

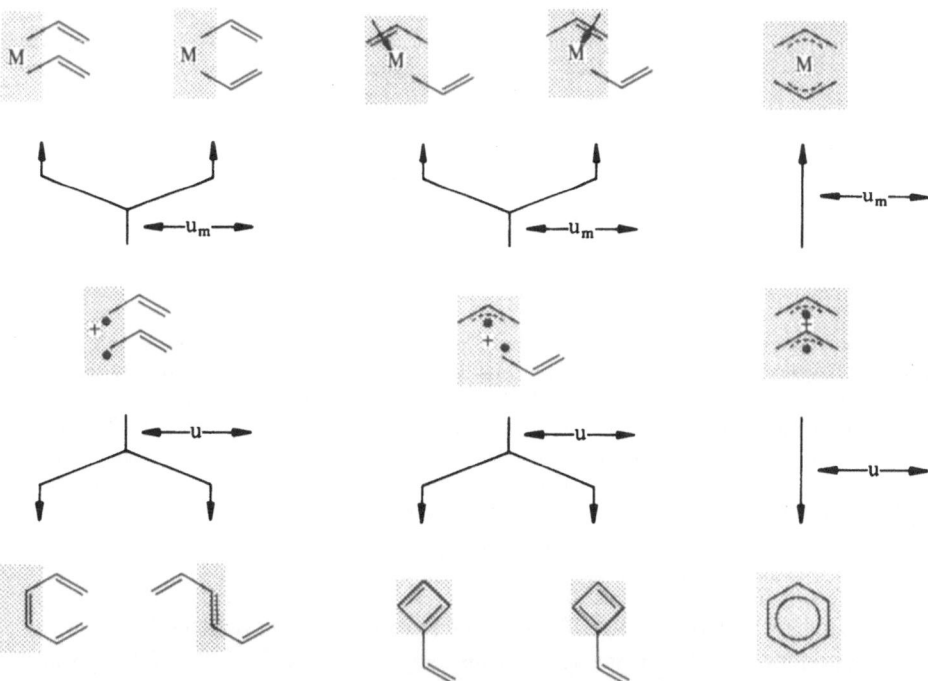

Scheme 2.5–1. Multifold unifying in organometallic and organic chemistry

Comparison of bis−σ−and bis−π−allyl metal complexes

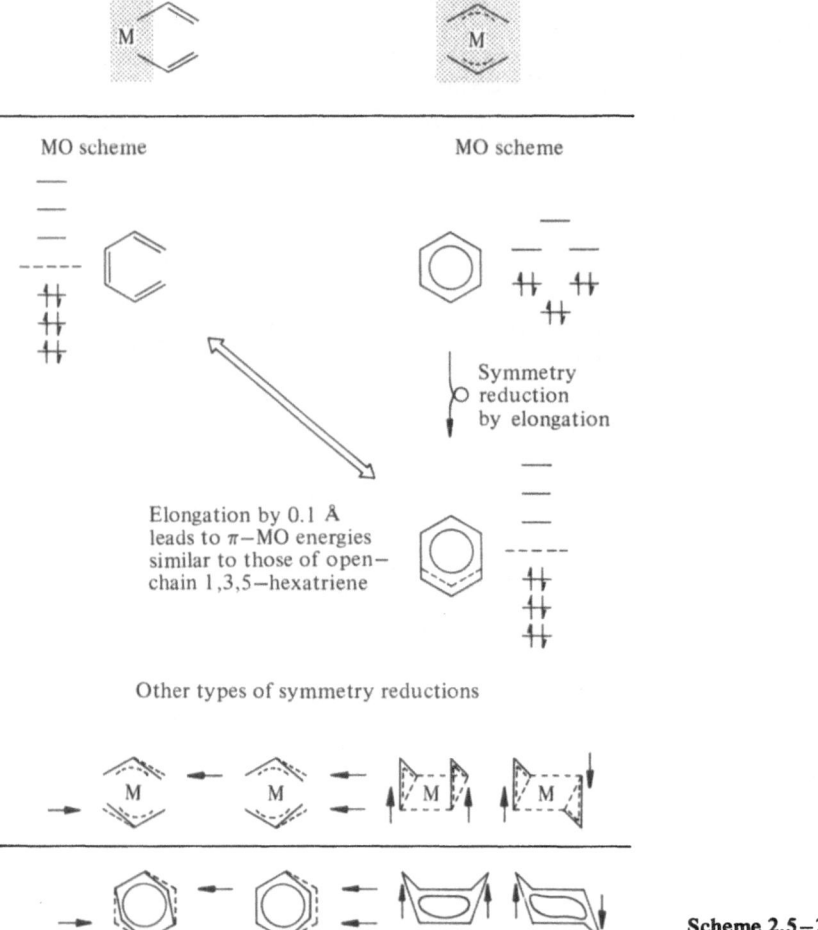

MO scheme   MO scheme

Symmetry
reduction
by elongation

Elongation by 0.1 Å
leads to π−MO energies
similar to those of open−
chain 1,3,5−hexatriene

Other types of symmetry reductions

Scheme 2.5−2

siders the symmetry reduction within the complexes. The combined bis-π-allyl system e.g. lacks the ideal D_{6h} symmetry of benzene. From X-ray analyses we know that, depending on the metal, the distance between the allyl groups varies. On the other hand, distortions and "chair"- or "boat"-forms could be found. Last not least, the allyl groups themselves are unsymmetric. The distance between the metal and the individual carbon atoms of the group and the degree of hybridization of the carbon atoms vary. This can be simulated by symmetry-relevant distortions within the skeleton of benzene as shown in Scheme 2.5-2.

How far can these analogies be applied to reaction control?

One contribution to this problem is the comparison of the PE spectrum of bis-π-allylnickel with that of 1,3,5-hexatriene. The upper part of Scheme 2.5-3 shows the interpretation of the PE spectrum of bis-π-allylnickel by *C. D. Batich* [86]. Important for our comparison is that the marked occupied orbitals of the unified ligand system (⊞) are ligand-centered and practically do not interact with metal orbitals.

Ligand—metal interaction scheme

For bis—π—allylnickel following the interpretation of PES data by C.D.Batich[86]

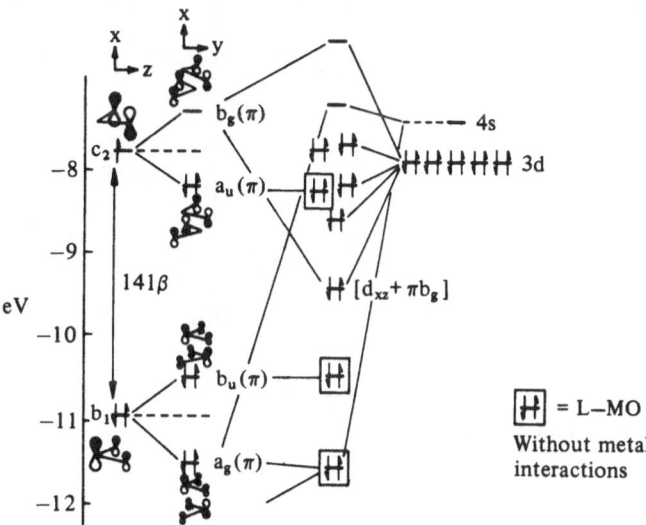

Comparison of direct— and metal—unified π—systems

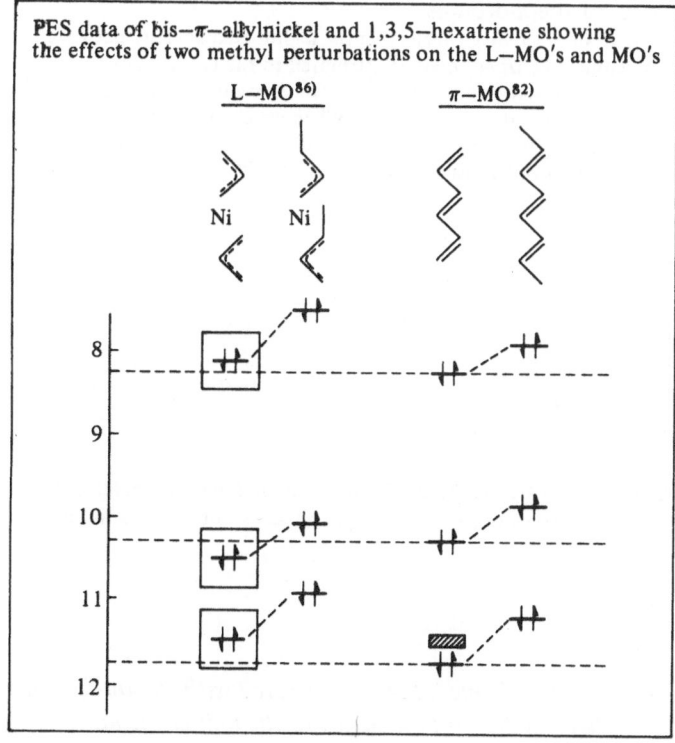

PES data of bis—π—allylnickel and 1,3,5—hexatriene showing the effects of two methyl perturbations on the L—MO's and MO's

L—MO[86] π—MO[82]

Scheme 2.5—3

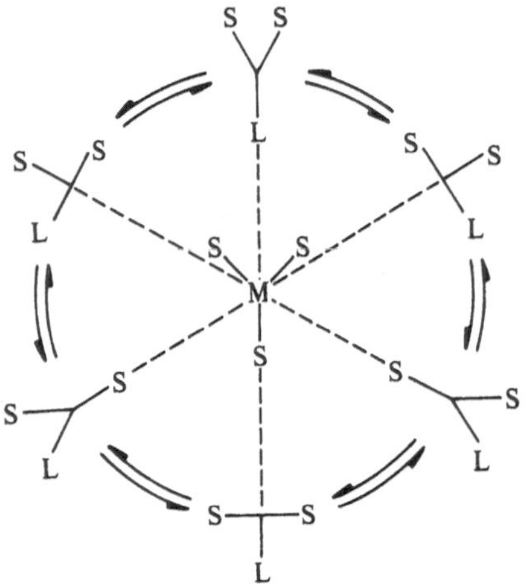

Fig. 1

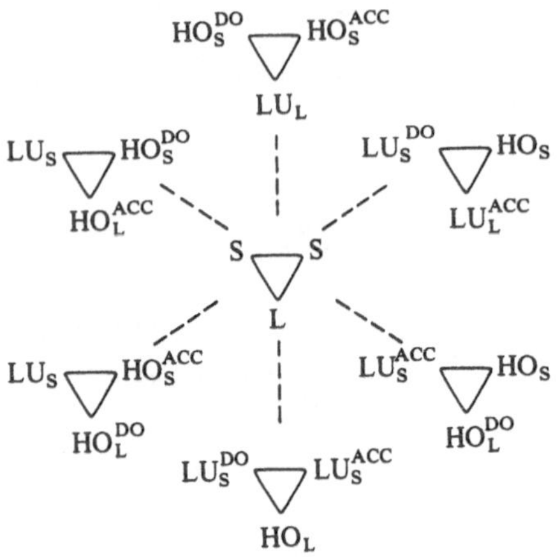

Fig. 2

Scheme 2.5–4. Structural and electronic situations

The ligand system is fixed to the metal by a single $HO_{metal}/LU_{unified\ allyls}$-interaction. At the bottom of Scheme 2.5-3 the occupied ligand-centered MO's of the metal-linked bis-allyl system are compared with 1,3,5-hexatriene-π-MO's. At the same time, the effect of a perturbation on both systems by two methyl groups clearly demonstrates the similarities in the energies of the π-orbitals.

That this comparison seems to have validity (neglecting the legitimated analogy between bis-π-allylnickel and symmetry-reduced benzene) may be attributed to the fact that open-chain π-systems may display some residual aromaticity (see Ref.[27]). Metal complexes with C_3 symmetry must have three identical ligands (S) in the same state. When we change only one ligand (S versus L) symmetry is reduced. The structural consequence is the change to a T- or Y-shaped moiety (see Fig. 1 in Scheme 2.5-4 and similar discussions in Refs.[83, 26b]). Within these arrange-

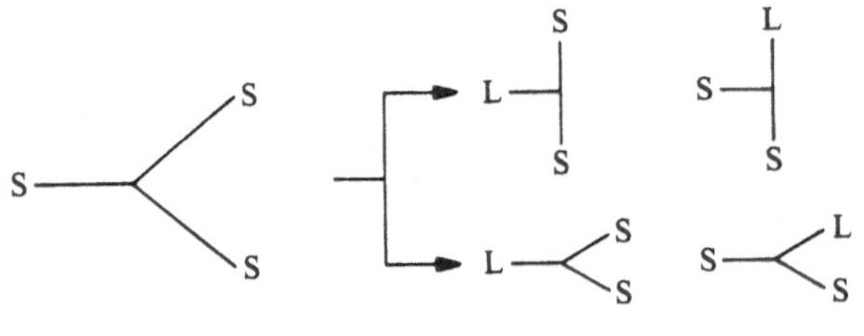

ments the L and S have HO/LU- and DO/ACC-functions (double-dual thinking see Scheme 2.3-1). According to the framework of the three-center FMO concept this leads to the six combinations shown in Fig. 2 of Scheme 2.5-4.

Question 2

How can the complexes in Fig. 1 of Scheme 2.5-4 be correlated with the three-center FMO schemes in Figure 2? What role has to be assigned to the unifier metal?

Answering these questions will be of pragmatic interest, too. In the ideal C_3 symmetry we have two important, degenerated d orbitals ($d_{x^2-y^2}$ and d_{xy}). Lowering the C_3 symmetry, this degeneracy is given up in a twofold way.

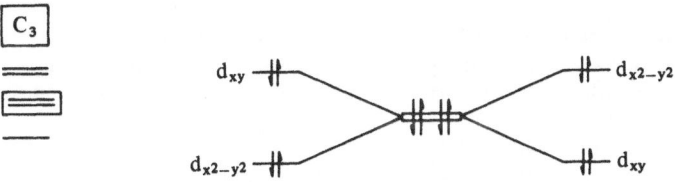

Variation within the unifier (e.g. additional ligand attack) or changes within the ligand field (spontaneous rearrangements) have different consequences depending on whether $d_{x^2-y^2}$ or d_{xy} contribute to the relevant orbital. This should lead to a predictible different behaviour in C–C-bond-formation and -splitting, ligand exchange or e.g. further association.

Question 3

Are there processes in metalorganic elementary steps, in which we have to consider dual reaction pathways, retaining and changing the conformation or configuration?

Example 1 (Ref.[117]):

Example 2 (see Scheme 2.3–5)

distribution		
–OCH$_3$	75%	25%
–SCH$_3$	30%	70%

Example 3 (Ref.[57]):

Example 4 (Ref.[5]):

Scheme 2.5–5.
Processes with change of configuration

75

There is ample preparative evidence that we have to assume dual reaction pathways which differ with respect to the symmetry behavior of the process. Here we are not faced with large differentiation energies leading to a concept of forbidden to allowed but with reactions separated by smaller activation energy differences more appropriate represented by the terms "preferred" and/or "restricted".

So, starting from syn-allyl complexes, cis- and trans-double bonds are generated depending on the type of perturbation (Scheme 2.5-5, examples 1–3). This type of information is stored within the complexes as isomers with reduced symmetry. This kind of symmetry reduction can be proved by direct spectroscopic investigations of the starting complexes (see e.g. in Scheme 2.3-5 NMR spectra of OCH_3- vs. SCH_3-complexes: double-dual differentiation of the isomers by O/S-exchange). Examples out of catalytic oligo- and polymerization [57] processes can best be interpreted by the same assumptions. Besides the O/S- or Ni/Pd-exchange, tipover in ACC/DO-P ligands has so far revealed this effect. The correlation diagram below shows the consequences within the MO picture (A/S processes[33]).

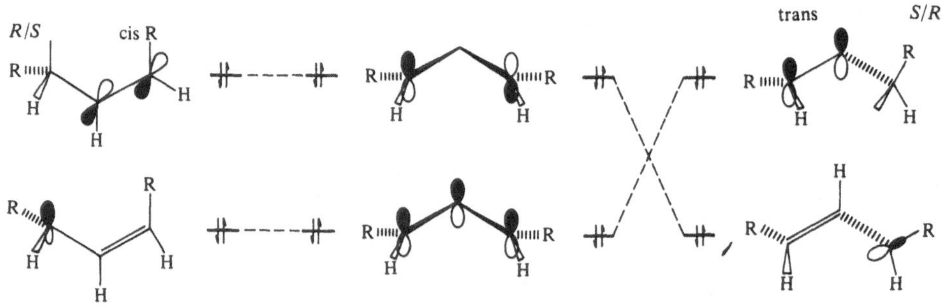

For dienes, the analogous correlation scheme can be established. This viewpoint leads for s-cis- and s-trans-butadiene to the following, preparatively utilizable consequences:

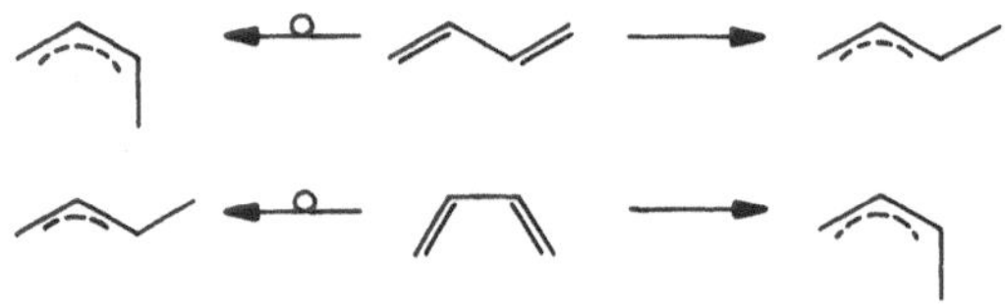

With respect to the problem of optical induction one has to keep in mind that cis- and trans-olefines are diastereomers.

3 Evidence for Ligand-Concentration Control by a "Titration of the Catalytic System" with Lewis Bases[20]

3.1 Introduction

Ligands having a lone electron pair on the phosphorus atom play a key role in transition-metal coordination chemistry of today. The range of complexes that can be produced by phosphorus donors is unparalleled in coordination chemistry. They are capable of stabilizing different coordination numbers, very high and low oxidation states (like Ni(0), Ni(I), Ni(II), Ni(III), Ni(IV) [87]), different complex geometries having the same stoichiometry (allogones) [88] and of inducing conformational and configurational changes in the organic moiety of the transition metal complexes under consideration (see Sect. 2.4).

The manifold intermediates in homogeneous transition-metal catalysis are certainly metal complexes and therefore show a behaviour like ordinary coordination compounds; associations of phosphorus donors open up multifarious additional controls. Both, substrates and P ligands are Lewis bases that we have to consider and that compete at the coordination centers of the metal, leading to competitive, non-competitive or uncompetitive activation or inhibition processes in analogy to the terminology of enzyme chemistry [89].

These effects are well-known in catalysis: Lewis bases are controlling factors in common use in technical processes like oxo process [90], Ziegler-Natta polymerization [91] and in those reactions depicted in Schemes 1.1-1.3 (Chap. 1). The actual number of ligands within the many intermediates and the induced electronic and steric ligand-property control within these complexes play an important role in the ligand-induced reaction control. The distribution of the catalytic active metal among the intermediate complexes in the given steady state is another important question of controlling catalysis. Nevertheless, these important aspects are still in a state

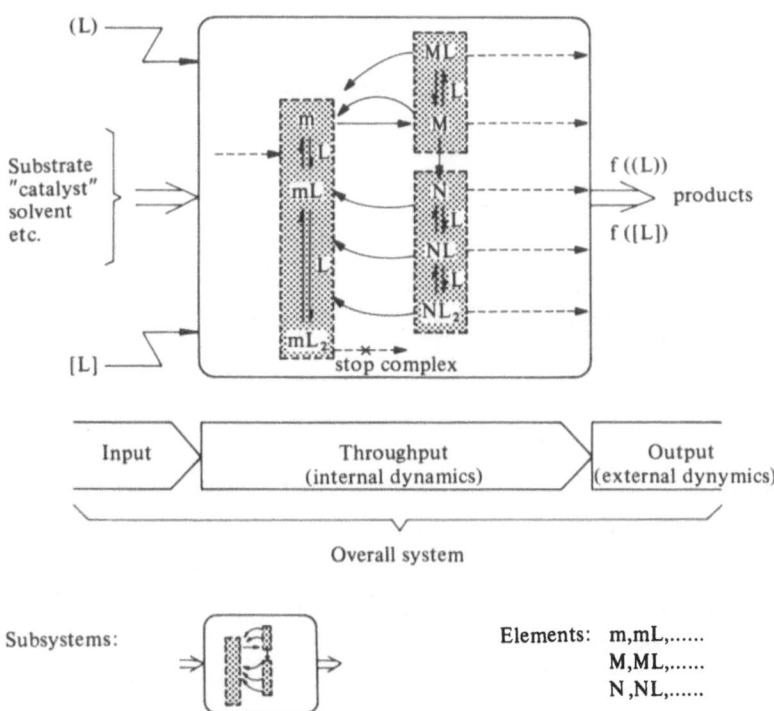

Scheme 3.1–1. Proposed structure of the overall system "Homogeneous Catalysis"

where "opinions may be sold as hard currency". This is due to the fact that up to now there has been no easily and widely applicable analytical method available of exploring these questions.

In the analysis of the structure of a system of high complexity, the way of thinking of the "systems analysis" has "organizing power". Accordingly, the catalytic system is like a "black box" containing intermediate complexes constituting the elements (see Scheme 3.1-1). These intermediates are related to one another in at least two ways. Intermediates being in direct equilibrium can be summarized to subsystems. Elements of two different subsystems are separated by kinetically controlled reactions and therefore in a lower degree of relationship than those of the same subsystem. Thus, we expect a hierarchic structure of the catalytic system. In contrast to spectroscopic methods, the complete inner structure of the system (the internal dynamics of the throughput) is not generally open to a direct experiment [21]. The method of choice is therefore the analysis of an "input-output relationship" of the overall system, the input containing the metal-starting complex, substrate, solvent etc, the output the resulting products, solvent etc.

A variation of the initial conditions of the input is related to a regular change of the product distribution. This change of the dynamics of the overall system allows an

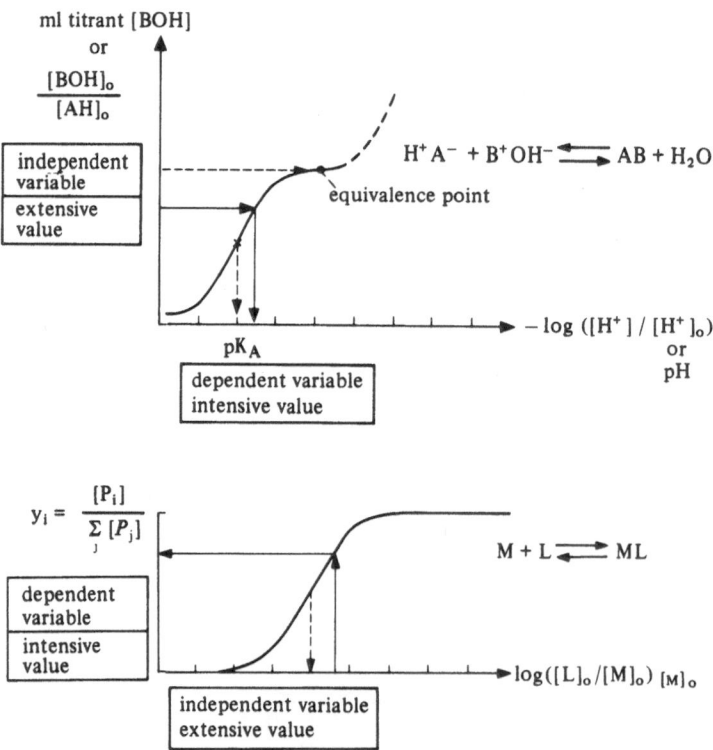

Scheme 3.1–2. Comparison of a diagram of a classical acid-base titration in aqueous medium with a [L]-control map for the "titration of the catalytic process"

insight into the inner structure of the "black box" homogeneous catalysis. Therefore, the varying product distribution (the selectivity) is plotted against the logarithmic ratio of the weighted amount of the controlling ligand and the metal complex $[\log ([L]_0/[M]_0)]$. The resulting curves have similarities with buffer curves in the classical acid-base titrations [92]. Examples for such diagrams are e.g. known in polymerization [93] and oligomerization [94] of olefins and even in competitive consecutive reactions in organic chemistry [95]. This type of "titration of homogeneous metal-catalyzed processes" is somewhat inverse to the classical titration in aqueous medium (Scheme 3.1-2).

The two variables change their role with respect to their dependent versus independent, intensive versus extensive nature. This is also true of e.g. calorimetric, conductometric and spectrophotometric titrations using UV-, IR- or NMR-spectroscopy [21, 96]. We additionally have to consider that in the "titration of the catalytic process" only the external dynamics are measured; a direct comparison with the actual metal fraction of the related intermediate complexes is generally not possible [97]. We call this analysis of homogeneous catalytic systems by a metal-ligand titration "the method of inverse titration" and for the resulting diagrams we use the term "ligand-concentration control maps ([L]-control maps)".

3.2 Ligand-Concentration Control Maps: Experimental Procedure – Analysis of the Resulting Diagrams [19]

For analyzing a catalytical system by the "method of inverse titration" one has to investigate at least 3 to 4 catalytic reactions for each chosen power of a tenth of the external metal-to-ligand ratio. To achieve this information efficiently in time, we carried out a special experimental procedure on the 1-ml scale (for experimental details see Ref. [98]), a flow chart is given in Scheme 3.2-1.

Standard solutions of the controlling ligand and the used metal complex are prepared under an inert gas atomosphere, both including internal standards (tetralin and n-dodecane, respectively) to determine gas chromatographically $\log ([L]_0/[Ni]_0)$ via the ratio of the two components. The ligand standard solution is then diluted in steps of powers of a tenth. Using the apparatus described in Fig. 3.2-1, the required quantities of the ligand and the metal solution are placed in glass tubes (Duran 50, $\phi_a = 8.0$ mm, $l = 27$ cm) and filled with solvent and substrate up to 1 ml. As an example, we used for the investigation of the catalytic system nickel/ligand/butadiene the ratio $1 : (10^{-6}$ up to $10^{1.5}) : 200$. The nickel concentration ($[Ni]_0$) was constant for the whole series of experiments. The tubes were then hermetically sealed under an inert gas atmosphere, placed into a water bath for the catalytic reaction, and then stored at -30 °C for GC analysis.

To prevent systematic mistakes in the dilution series of the ligand standard solutions, leading to relative shifts in the [L]-control maps, we carried out independent control catalyses on the 250-ml scale. For the ($[L]_0/[Ni]_0$) ratio we selected inflection points in the varying product distribution of the [L]-control maps. In Fig. 3.2-2 is exemplified the [L]-control map of the catalytic system nickel/phenyl-diphenoxi-phosphine/butadiene [99].

The seven products ($2 - 7$) are formed in varying amounts [100]. Depending on the ($[L]_0/[Ni]_0$) ratio, several association processes are recognizable. The controlling

P. Heimbach and H. Schenkluhn

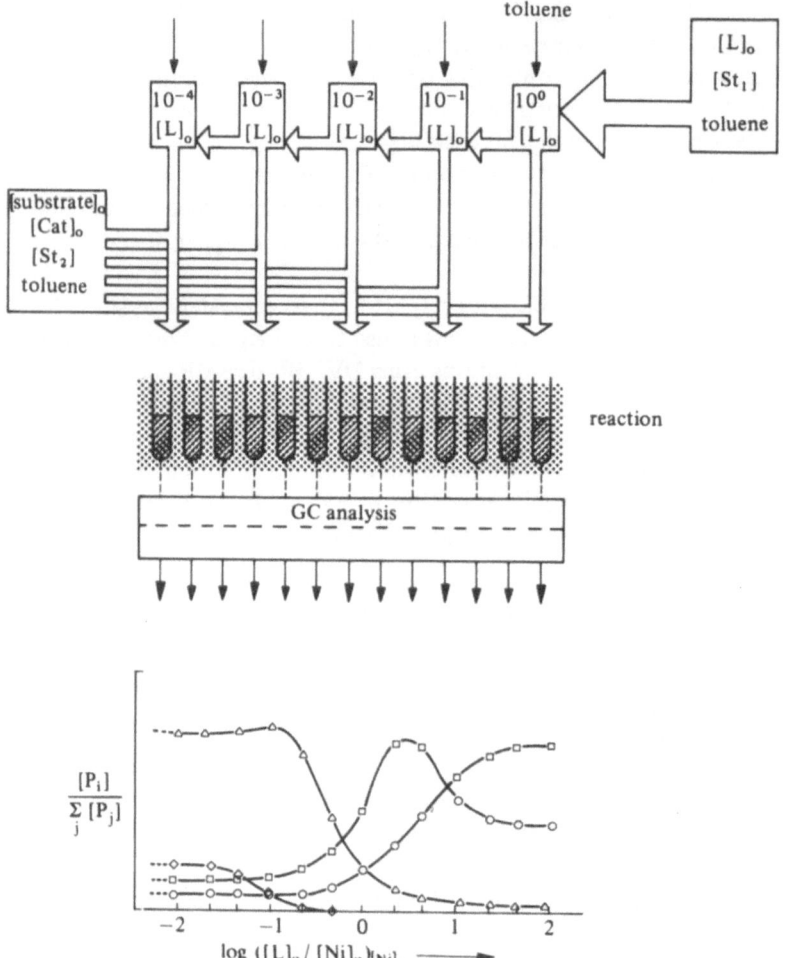

Scheme 3.2−1. Flow chart of the experimental procedure for analyzing a catalytic system via the "method of inverse titration"
St_1 = • first internal standard
St_2 = • second internal standard

Fig. 3.2−2. [L]-control map for the catalytic system COD_2Ni/diphenoxi-phenylphosphine/butadiene $(1:x:170)$[99]
Reaction conditions: $[Ni]_0$ = 40 mmol/l[118]), T = 60°, t = 15 h, solvent: toluene, conversion approx. 95%. a) [L]-control map; b) 15% detail-scale up; products: 2: *ttt* CDT, 3: *ttc* CDT, 4: *tcc* CDT, 5: COD, 6: VCH, 7: n-OT, educt: 1: butadiene. selectivity y_i in mol%

$$(y_i = [P_i]/ \sum_{j=1}^{7} [P_j])$$

▶

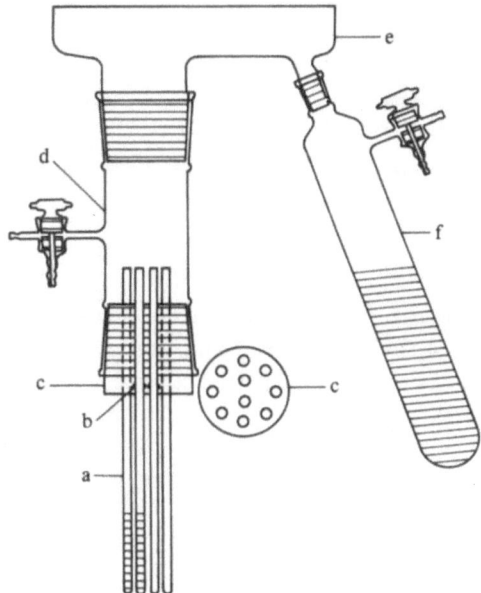

Fig. 3.2–1. Experimental build-up:
a) glass tube; b) boreholes with slip ring
seals; c) aluminium block NS 60; d) glass
head; e) inert gas bridge; f) ampoule with
a standard solution

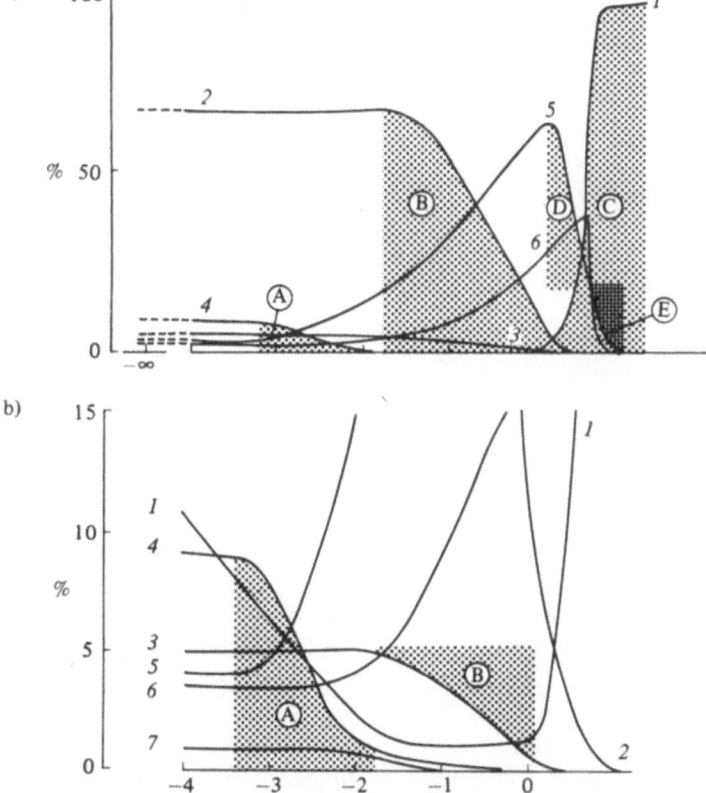

$$\log \left([L]_o \,/\, [Ni]_o \right)_{[Ni]_o}$$

P. Heimbach and H. Schenkluhn

first association of the ligand occurs at different ligand-to-metal ratios depending on the products.

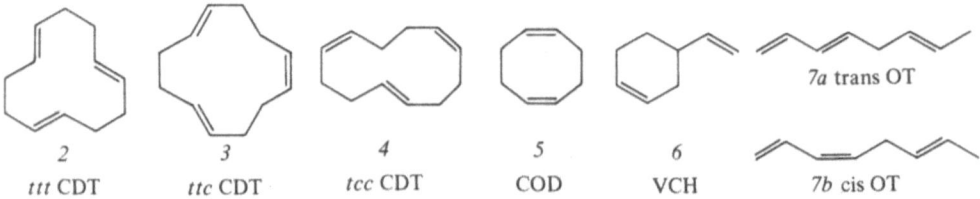

2	3	4	5	6	7a trans OT
ttt CDT	ttc CDT	tcc CDT	COD	VCH	7b cis OT

To obtain more information on how the nature and number of the elementary steps are influenced in situ by the addition of the controlling ligand, partial control maps were deduced from the original one. In Scheme 3.2-2 partial control maps for

Fig. 1. Oligomer map

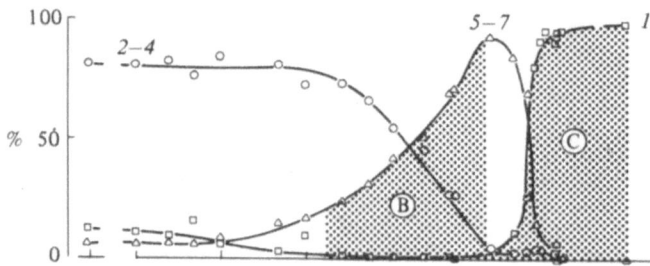

Fig. 2. Dimer map

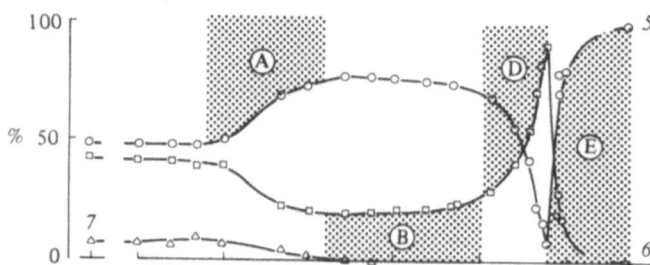

Fig. 3. Trimer map

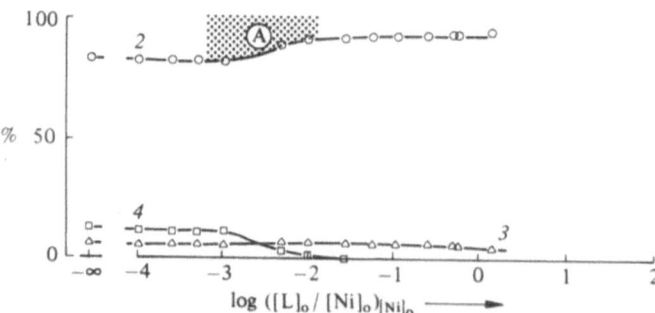

$$\log \left([L]_o / [Ni]_o \right)_{[Ni]_o} \longrightarrow$$

Scheme 3.2–2. Partial maps for the catalytic system COD_2Ni/phenyldiphenoxi-phenylphosphine/butadiene (1:x:170) (reaction conditions see Fig. 3.2–2)

(1) the degree of oligomerization, the distribution of (2) the dimers and (3) the trimers are shown.

In the oligomer map two almost typical titration curves (B, C) are obtained. Their positions on the log $([L]_0/[Ni]_0)$ scale indicate that the corresponding intermediates occur in high steady-state concentrations.

In the dimer map four association processes (A, B, D, E) are recognizable. The first one (A) indicates an association process for an intermediate at low steady-state concentration. Association process B corresponds to the first ligand control in the oligomer map.

The trimer map is the simplest one. One association process is visible as a typical titration curve. This corresponds to the association process A in the dimer map. An examination of the [L]-control map indicates an alteration in the speed of formation of the trimer (4) and the dimer (5) as the original change in the product pattern.

Comparing [L]-control maps for up to now fifteen different ligands of the nickel-catalyzed cyclooligomerization of butadiene we found eight independent product-determining ligand-association processes (I − VIII). To find out the respective changes in the [L]-control maps, it is extremely helpful to construct "product stream diagrams" as described in Scheme 3.2-3.

For each association phenomenon a resulting increase of one product is symbolized by a white button, a decrease by a black button. In addition, the direction of the change in the product distribution is marked by an arrow. From these eight controlling processes five can be localized in the presented [diphenoxi-phenyl phosphine]-control map (Fig. 3.2-2 and Scheme 3.2.-2).

A "preliminary rationale" for the steady-state situation of the catalytic system is given in Scheme 3.2-4. The product-controlling ligand association processes are particularly marked. Intermediates of the metal-olefin type are symbolized by S_nNi, n being two to four; intermediates containing a bis-allyl C_8-chain as organic ligand are symbolized by $H_4C_2{}^{A}_{A}$ Ni, those containing a bis-allyl C_{12}-chain by $H_{10}C_6{}^{A}_{A}$ Ni.

Ligand association processes

Products	I P	I C	II P	II C	III/A P	III/A C	IV P	IV C	V/B P	V/B C	VI/C P	VI/C C	VII/D P	VII/D C	VIII/E P	VIII/E C
2−4 TRIM	●		=		○		○		●		●		●			
5−7 DIM	=		=		○		○		○		●		●		●	
1 MON	○	○⟶	=	=	=	=	●	●⟶	=	=	○	○⟶	○	○	○	○
5 COD	=	(=)	=	=	○	○⟶	=	( )	○^A_D	○⟶	○	●	○^A_D	●	○^{A}	●
6 VCH	=	(=)	=	=	●	=	=	( )	●	○⟶	○	●	●	●	●	●
7 n−OT	=	(=)	=	=	( )	( )	( )	( )	( )	( )	( )	( )	( )	( )	( )	( )
2 ttt CDT	=	●⟶	○	○⟶	=	●⟶	○	○	=	●⟶	=	●⟶	=	●		
3 ttc CDT	=	●⟶	=	=	=	●⟶	=	●⟶	=	●⟶	=	●⟶	=	●		
4 tcc CDT	=	●⟶	●	●⟶	=	( )	( )	○⟶	=	●⟶	=	●⟶	=	●		

Scheme 3.2−3. "Product-stream diagram" for the catalytic system COD_2Ni/Lewis base/butadiene.

○ increase of the fraction of the designated product
● decrease of the fraction of the designated product
= indifferent behaviour
( ) not yet determined

A acceptor-Lewis base (HO_L^{ACC})
D donor-Lewis base (HO_L^{DO})
P partial map
C complete map

I−VIII ligand association processes
A−E ligand association processes referring to Fig. 3.2−2 and Scheme 3.2−2

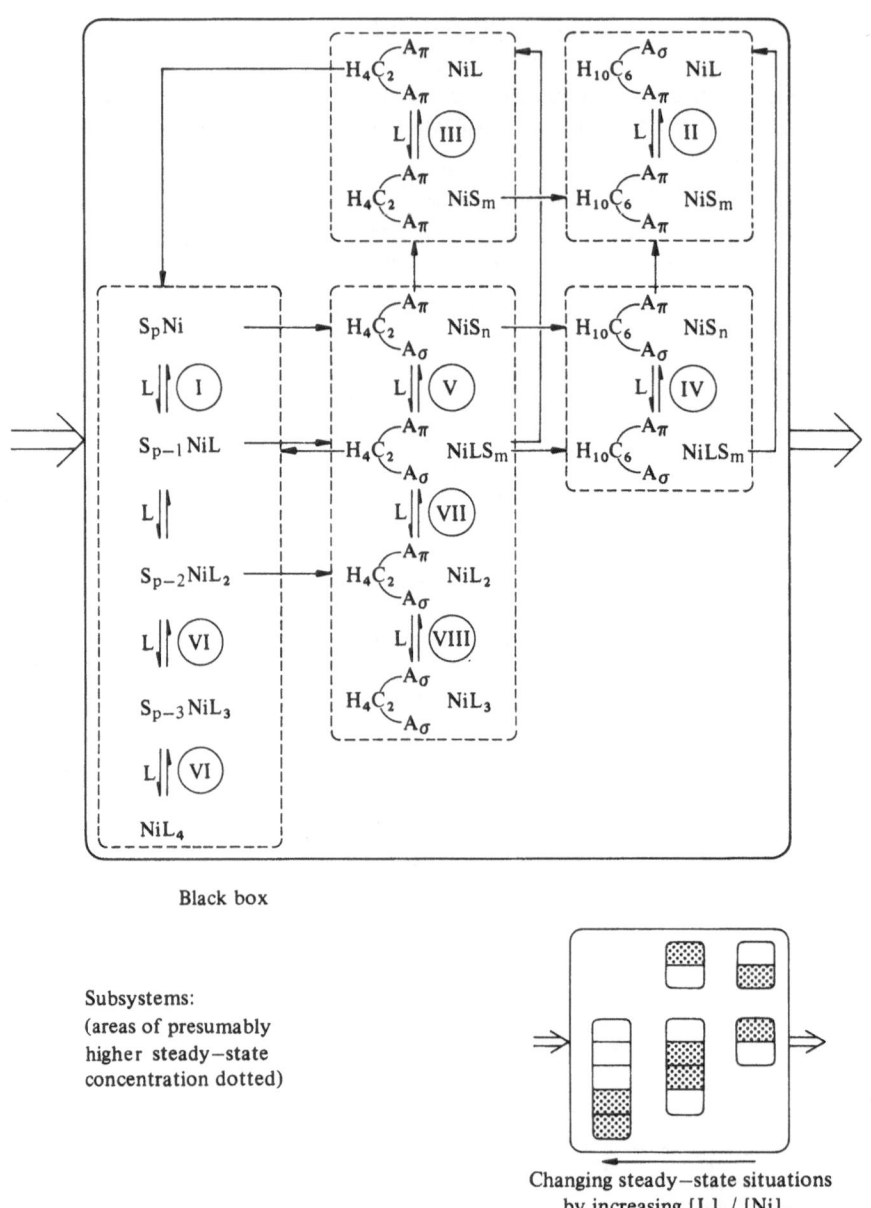

Black box

Subsystems:
(areas of presumably
higher steady—state
concentration dotted)

Changing steady—state situations
by increasing $[L]_o / [Ni]_o$

Scheme 3.2–4. "Preliminary rationale" for the steady state situation of the catalytic system COD_2Ni/Lewis base/butadiene m = 0 or 1, n = 1 or 2, p = 2 to 4, S = substrate (butadiene)

Some of the assumed intermediate complexes have already been isolated and characterized.

We have chosen these schematic formulas, because the occurrence of syn/anti- and cis/trans-isomers gives rise to new problems, not discussed in detail in this publication (see Sect. 2.5).

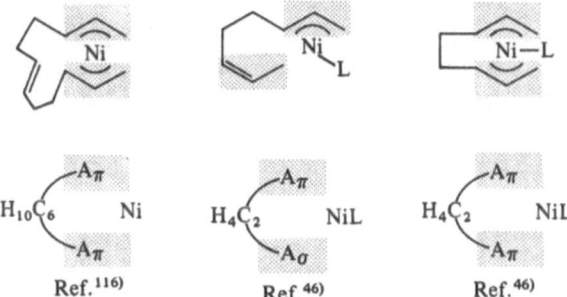

Ref.[116]   Ref.[46]   Ref.[46]

As a bridge between the original [L]-control maps, the "product stream diagrams" and the rationale concerning the steady-state situation of the intermediate complexes, we use "buffer curve diagrams", which symbolize the range of the existence for the many intermediate complexes on the log $([L]_0/[Ni]_0)$ scale. Scheme 3.2-5 describes the "buffer-curve-diagram" for the [diphenoxi-phenyl phosphine]-control map, the intermediates in higher steady-state concentration being denoted by shaded areas.

As we have shown, the discussion of the product curves, in analogy to classical titration curves, affords information on the number of the intermediate complexes, their relation to each other and their steady-state concentration leading to a preliminary reaction scheme.

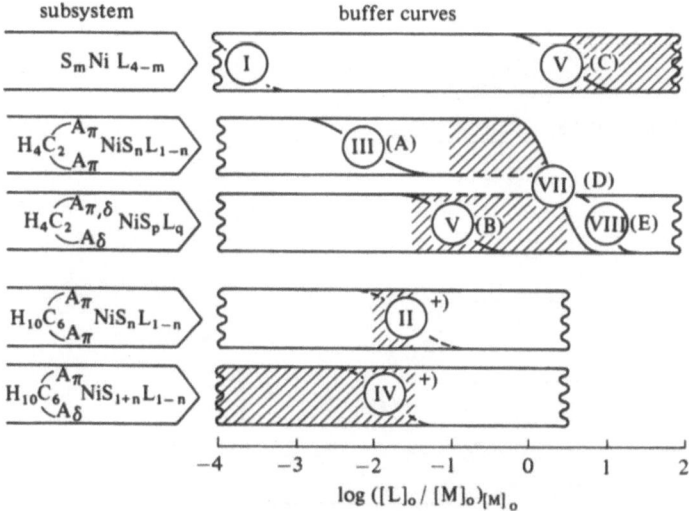

Scheme 3.2–5. "Buffer curve diagram": range of existence for the many intermediates classified into subsystems. Solid lines: buffer curves for the system COD_2Ni/diphenoxi-phenylphosphine/ butadiene. Broken lines: remaining curves for this system (only supposed, but demonstrable for other Lewis bases). Areas with higher steady state concentration shaded. (m = 2 to 4, n = 0 or 1, p + q = 2 or 3, p = 0, 1, 2 or 3)

a   Steady-state situation is not yet conclusive. At low temperature only $H_{10}C_6$ ⟨A_π / A_π⟩ Ni can be isolated.

85

For many purposes, this may be sufficient and needs no further investigations. For a complete picture, a more detailed analysis of the many elementary steps within the coexisting catalytic cycles is necessary; for their range of existence, the initial conditions can be defined by [L]-control maps [101]. An example for this is given in Sect. 3.4.

Fig. 1. System $Ni(COD)_2$/triphenylphosphine/butadiene $(1:x:170)$[102]

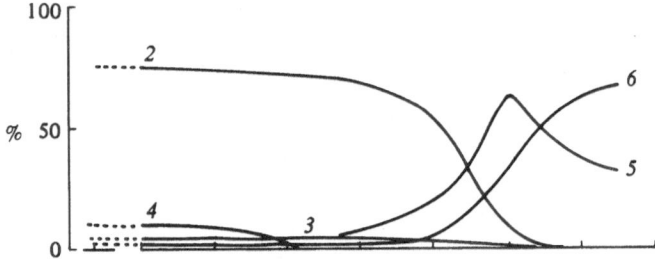

Fig. 2. System $Ni(COD)_2$/triphenylphosphine/butadiene/propene
$(1:x:200:200)$[22,103]

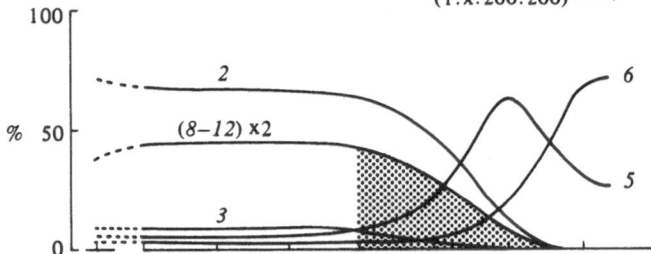

Co—oligomer map

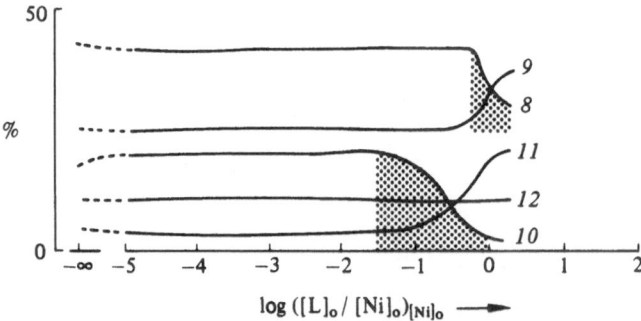

$$\log\left([L]_o / [Ni]_o\right)_{[Ni]_o} \longrightarrow$$

Scheme 3.3–1. [L]-control maps for a three- and a four-component system (product distribution in mol%)
Fig. 1: reaction conditions see Fig. 3.2–2.
Fig. 2: reaction conditions: $[Ni]_0 = 40$ mmol/l, T = 40 °C, t = 144 h, solvent: toluene; butadiene conversion approx. 95%

3.3 Application of the Ligand-Concentration Control Maps in Transition Metal Chemistry

The [L]-control maps can be used not only for the first analysis of the mechanism (minimum number of intermediate complexes, their product-determining manifold ligand association processes and their coupling) of homogeneous metal-catalyzed reactions but also for the expansion of catalytic systems to four-, five- or even six-component systems. The role of the new component can in many cases be easily deduced from the changes of the pattern of the corresponding [L]-control maps.

In Scheme 3.3-1 the [L]-control maps for the three-component system nickel/triphenylphosphine/butadiene [102] and the four-component system nickel/triphenylphosphine/butadiene/propene [22, 103] are compared. The amount of the produced 2:1 co-oligomers of butadiene and propene are strongly proportional to the trimers of the butadiene formed. This suggests a subsystem, in which the trimerization and dimerisation of butadiene and co-oligomerisation of butadiene and propene are commonly governed by a thermodynamically controlled competition of butadiene, propene and the directing ligand. The co-oligomer map, i.e. partial map of the distribution of the co-oligomers as a function of $\log ([L]_0/[Ni]_0)$, indicates that two ligand-association processes determine the ratio of cyclic to open-chain products. Ligand association favors the formation of cyclic products. The existence of the two titration curves suggests at least two kinetically separated subsystems, the one forming the co-oligomers (8) and (9) and the other the co-oligomers (10) and (11).

The invaribility of co-oligomer (*12*) against a variation of the ($[L]_0/[M]_0$) ratio suggests an independent way for its generation. A preliminary scheme for the additionally formed intermediate complexes of the co-oligomerization is given on page 87.

In the system nickel/L/butadiene, secondary amines can shift the cyclodimerization of butadiene to the acyclic products (*7a*) and (*7b*)[11, 104]. Its cocatalyst function can be visualized by the corresponding [L]-control map (Scheme 3.3-2). In the three-component system nickel/morpholine/butadiene[12] the open-chain products are formed for log ($[morpholine]_0/[Ni]_0$) > -1. Both octatrienes (*7a*) and (*7b*) are formed at the constant ratio of 1.8 over the entire range of the examined amine/nickel scale. However, the efficiency of the catalytic system is low. After a turnover of 30% butadiene, the catalytic activity ends because of the formation of stop complexes of the nickel amide type.

Fig. 1. System Ni(COD)$_2$/morpholine/butadiene (1:x:120)[12]

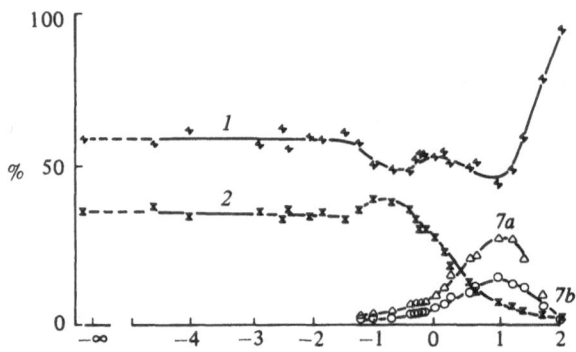

Fig. 2. System Ni(COD)$_2$/triphenyl phosphite/morpholine/butadiene (1:x:10:120)[12]

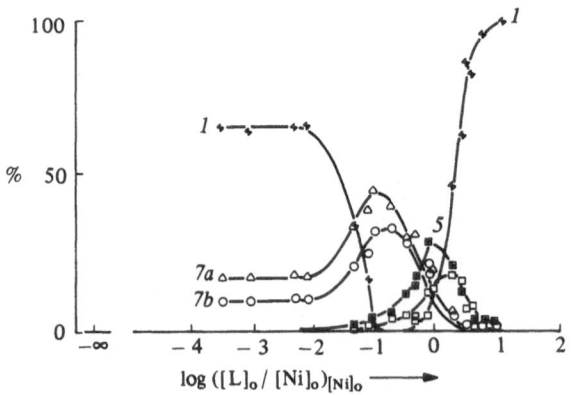

log ($[L]_0/[Ni]_0)_{[Ni]_0}$ ⟶

Scheme 3.3–2. [L]-control maps for the oligomerization of butadiene (educt-product distribution in mass%). Reaction conditions: $[Ni]_0 = 40$ mmol/l[118], T = 40 °C, t = 16 h, solvent: toluene. + *1:* butadiene, × *2: ttt* CDT, ▪ *5:* COD, □ *6:* VCH, △ *7a:* trans OT. ○ *7b:* cis OT

We now turn to the four-component system nickel/triphenyl phosphite/morpholine/butadiene $(1:x:10:120)$ [12]. At low [triphenyl phosphite]$_0$/[nickel]$_0$ ratio $(= 10^{-2})$ both ligands act co-operatively. The presence of the additional P-ligand prevents from the stop complex formation. An intermediate complex of low steady-state concentration plays the key role. At higher [P-ligand]$_0$/[nickel]$_0$ ratios (from 10^{-1}) both Lewis bases act competitively, the amine as directing ligand leading to the open-chain products (7a) and (7b), the P-ligand leading to the cyclodimers (5) and (6).

This example impressively demonstrates the pragmatic value of [L]-control maps for the discovery of the best catalytic "channels" for a desired product. We successfully used this method applying optimal reaction conditions in the preparation of five- and six-component systems[111].

Fig. 1. System Ni (COD)$_2$ /Al (Et)Cl$_2$ /tricyclohexylphosphine/propene
$(1:4:x:200)$

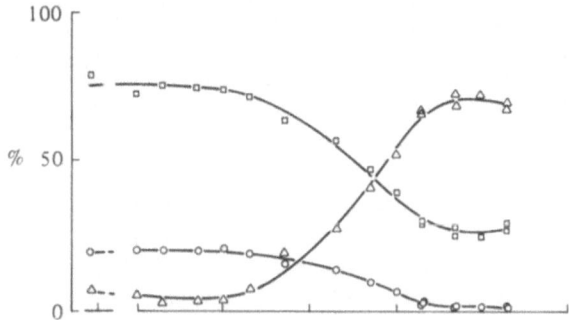

Fig. 2. System (allyl) NiCl/Al (Et)Cl$_2$ /tricyclohexylphosphine/propene

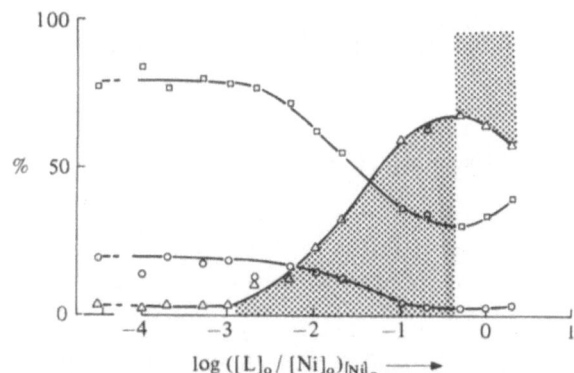

$$\log([L]_o / [Ni]_o)_{[Ni]_o}$$

Scheme 3.3–3. Propene dimerization using homogeneous nickel catalysts (product distribution in mol%)[109]. Reaction conditions: $[Ni]_0 : [Al (Et) Cl_2]_0 : [propene]_0 = 1:4:200$, $[Ni]_0 = 20$ mmol/l, $T = -25\,°C$, $t = 4$ h, solvent: chlorobenzene, conversion approx. 50%. ○ hexenes, □ methylpentenes, △ 2,3-dimethylbutenes

Another important application of [L]-control maps is the comparison of catalyzed processes. For the dimerization of propene at homogeneous nickel catalysts, two types of starting complexes are used, nickel-olefin complexes like bis-1,5-cyclooctadiene-nickel[107] and allyl-halo-nickel complexes like bis-(π-allyl-chloro-nickel)[108]. The latter leads to a more active catalyst; using the former, an incubation time after mixing of the components has to be taken into account. We were interested in the question wether the resulting catalytic systems are comparable or not. The dimers formed are of three different skeletons, hexenes, methyl-pentenes and 2,3-dimethyl-butenes. We investigated the catalytic systems Ni $(COD)_2$/AlEtCl$_2$/tricyclohexylphosphine/propene $(1:4:x:200)$ and allyl-NiCl/AlEtCl$_2$/tricyclohexylphosphine/propene $(1:4:x:200)$[109]. The resulting [L]-control maps are shown in Scheme 3.3-3.

The dimers are listed under the three possible skeletons ignoring the isomerization properties of the two catalytic systems. As it can easily be seen there is no difference in the skeleton-determining step between both systems from the standpoint of the [L]-control maps. One product-determining ligand-association process is recognizable, the range of existence and the changing product pattern being the same for both starting complexes.

Up to now, on the abscissae of all presented [L]-control maps, the logarithm of the external ligand/metal ratio with constant metal concentration (log $([L]_0/[M]_0)[M]_{0 = \text{const.}}$) has been plotted. Therefore, these [L]-control maps were established by varying the total concentration of the directing ligand ($[L]_0$) at an invariable total concentration of the metal component ($[M]_0$). In general, two other types of systematic variations of the ligand/metal ratio are possible, the variation of the total concentration of the metal ($[M]_0$) at constant external ligand/metal ratio (abscissa: log $([M]_0)[L]_0/[M]_0 = \text{const.}$) and the variation of the total concentration of the metal ($[M]_0$) at constant total ligand concentration ($[L]_0$) (abscissa: log $([L]_0/[M]_0)[L]_0 = \text{const.}$).

In Scheme 3.3-4 these three possible types of [L]-control maps for the catalytic three-component system nickel/diphenyl-phenoxiphosphine/butadiene are shown[110, 111]. Only the curves of the major products are presented: cyclotrimer *ttt* CDT (2) and both cyclodimers COD (5) and VCH (6). The resulting patterns indicate these three types of variations of the ligand to metal ratio as being somewhat related to one another. A single association process of the directing ligand (association process B in Scheme 3.2-2) determines the degree of oligomerization, the degree of association α being the relevant parameter for this controlling process. The parameter α is a function of the mass action law as well as of Ostwald's dilution law[92]. Therefore, it depends on the dissociation constant K_D and the total concentration of the ligand associate.

$$M + L \overset{K_D}{\rightleftharpoons} ML \qquad \alpha = \frac{[ML]_0 - [ML]}{[ML]_0} = f\,(K_D, [M]_0, [L]_0)$$

$$K_D = \frac{[M] \cdot [L]}{[ML]}$$

These three types of [L]-control maps strikingly remind us of the important role of Ostwald's dilution law in homogeneous catalysis.

Case I (selectivity $y_i = P_i/\Sigma P_i = f\,([L]_0/[M]_0)[M]_0 = \text{const.}$) describes the ligand control in catalytic systems using closed reactors (autoclaves, ampoules etc.) $[M]_0$ does not change during the reaction; the product distribution is determined by the ligand to metal ratio.

Case II (selectivity $y_i = P_i/\Sigma_{P_i} = f\,([M]_0)[L]_0/[M]_0 = const.$) is realized in catalytic systems using open reactors whereas the catalyst is applied right from the beginning and substrate is permanently delivered. For highly active systems, the volume of the reaction mixture often expands throughout this procedure by two to three powers of a tenth resulting in a decrease of the degree of association for the intermediate complexes. Its influence on selectivity can be calculated by this type of [L]-control maps. Moreover, these maps give an idea of the degree of selectivity

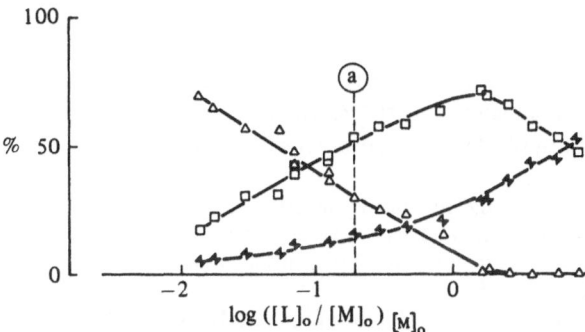

Case I: $[M]_0 = 0.04$ mmol/l[118]

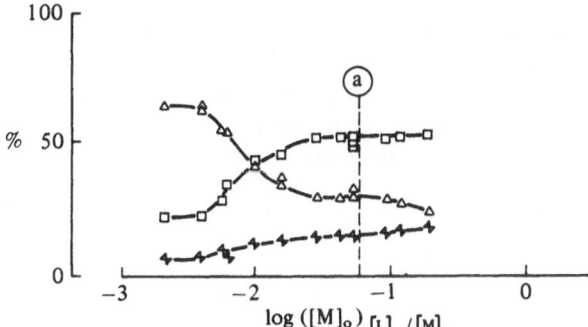

Case II: $[L]_0/[M]_0 = 0.2$

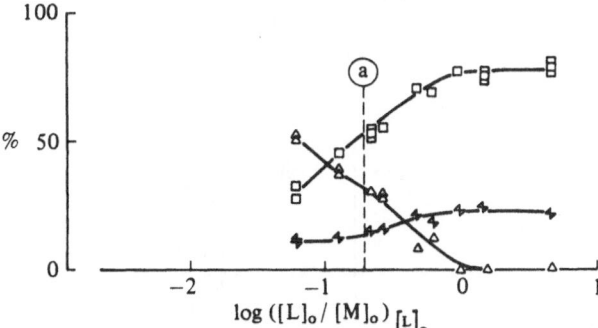

Case III: $[L]_0 = 0.008$ mmol/l[118]

Scheme 3.3–4. Catalytic system COD_2Ni/diphenyl-phenoxiphosphine/butadiene[111] (product distribution in mol%).
Reference reaction Ⓐ $[Ni]_0:[L]_0:[butadiene]_0 = 1:0.2:160$, reaction conditions see Fig. 3.2–2

of a given catalytic system with respect to concentration profiles in a flow reactor. Finally, these results provide a simple explanation for the fact that for extremely active catalysts used in the ppm range (like rhodium catalysts in the hydroformylation reaction), ligand/metal ratios of up to 10^5 are needed in order to promote the desired (two) product-determining ligand association processes.

Case III (selectivity $y_i = P_i/\Sigma P_i = f ([L]_0/[M]_0)_{[L]_0 = const.}$) is realized in open reactors where the catalyst is used right from the beginning and substrate as well as the controlling ligand are permanently delivered at a constant ratio. This sometimes is the best strategy to overcome the influence of Ostwald's dilution law in catalysis, but — as Scheme 3.3-4. indicates — fails below a critical $([L]_0/[M]_0)_{[L]_0 = const.}$ ratio.

As Scheme 3.3-5 demonstrates, the application of [L]-control maps to reactions in coordination chemistry leads to analogous results [40]. The fraction of the CO-insertion product in the decomposition of (13) in the presence of CO typically depends on the ligand to metal ratio.

At $-78\,^\circ$C, 0.2 molar solutions of 13 and the directing ligand ethoxi-diphenyl-phosphine were poured together under an inert gas atmosphere. After 5 minutes excess CO was added. The temperature was then allowed to raise to room temperature. The resulting ratio of the organic products was determined gas chromatographically.

The absence of a directing influence of the controlling ligand below the first association process at $\log ([L]_0/[M]_0) = 0$ indicates that this decomposition proceeds stoichiometrically rather than catalytically. (In contrast, the [L]-control map for the decomposition of 13 in the absence of CO opens up additional catalytic pathways [53]). For this reaction, the range of existence for the ligand associates of the educt- and product-complexes are determined by NMR spectroscopy indicating changes in the number of associated ligands from one subsystem to the next. The reaction equations for the ratios $([L]_0/[Ni]_0) = 1$ and 2 are given below.

$[L]_0 / [Ni]_0 = 1$

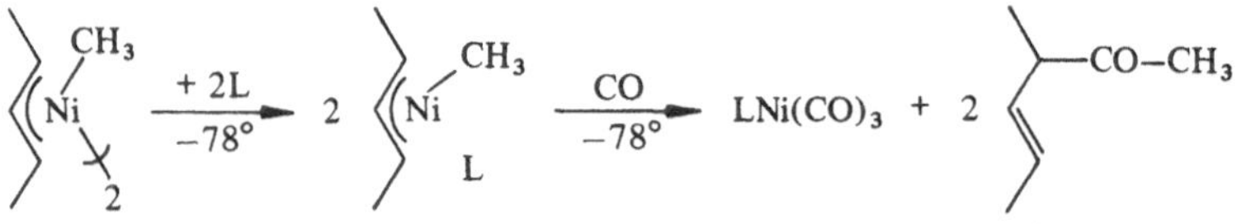

13   14   16

$[L]_0 / [Ni]_0 = 2$

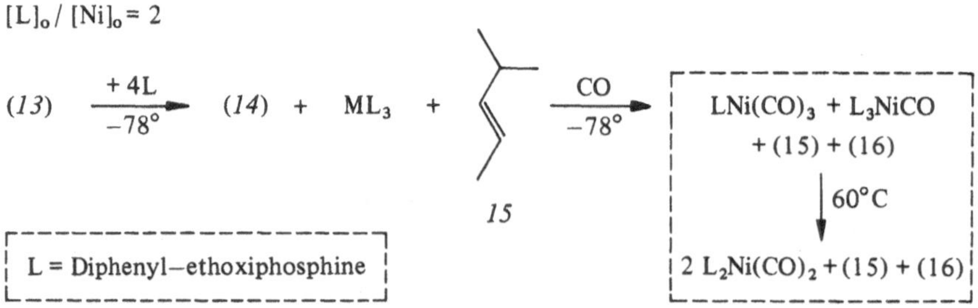

L = Diphenyl–ethoxiphosphine

Comparable results are known for the decomposition of nickela-cyclopentane complexes which lead, depending on the ligand-to-metal ratio to cyclobutane, 1-butene and ethylene [112].

Fig. 1. Reaction products of the system *13* / diphenyl–ethoxiphosphine / CO

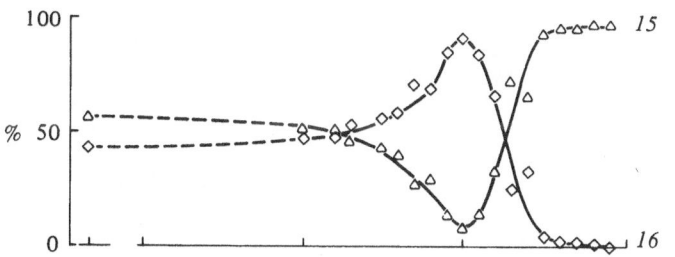

Fig. 2. Complex chemical system *13* / diphenyl–ethoxiphosphine at $-78°C$ before addition of CO

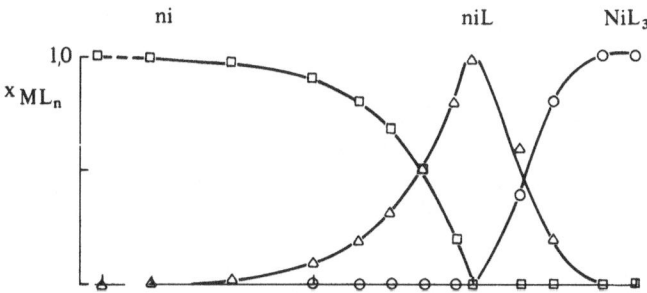

Fig. 3. Complex chemical system $Ni(CO)_4$ / diphenyl–ethoxiphosphine corresponding to the distribution after completed reaction at 60 °C

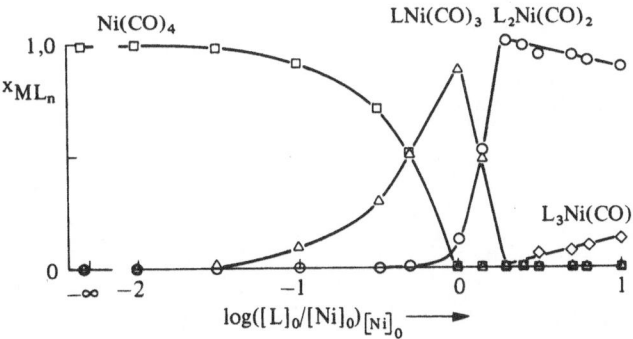

Scheme 3.3–5. [L]-control maps applied to coordination chemistry for investigating the chemical reaction

Figure 2: $K_1 \leqslant 10^{-4}; K_2 \geqslant 10^2; K_3 \leqslant 10^{-3}; K_4 \geqslant 10^4$ [Mol]

Figure 3: $K_1 \leqslant 5\,10^{-6}; K_2 \simeq 10^{-3}; K_3 \simeq 50; K_4 \geqslant 10^3$ [Mol]

Reaction conditions: $[Ni]_0 = 0.2$ mol/l, p_{CO} = standard pressure, T = -78 °C, solvent:toluene, ni = 0.5 (*13*)

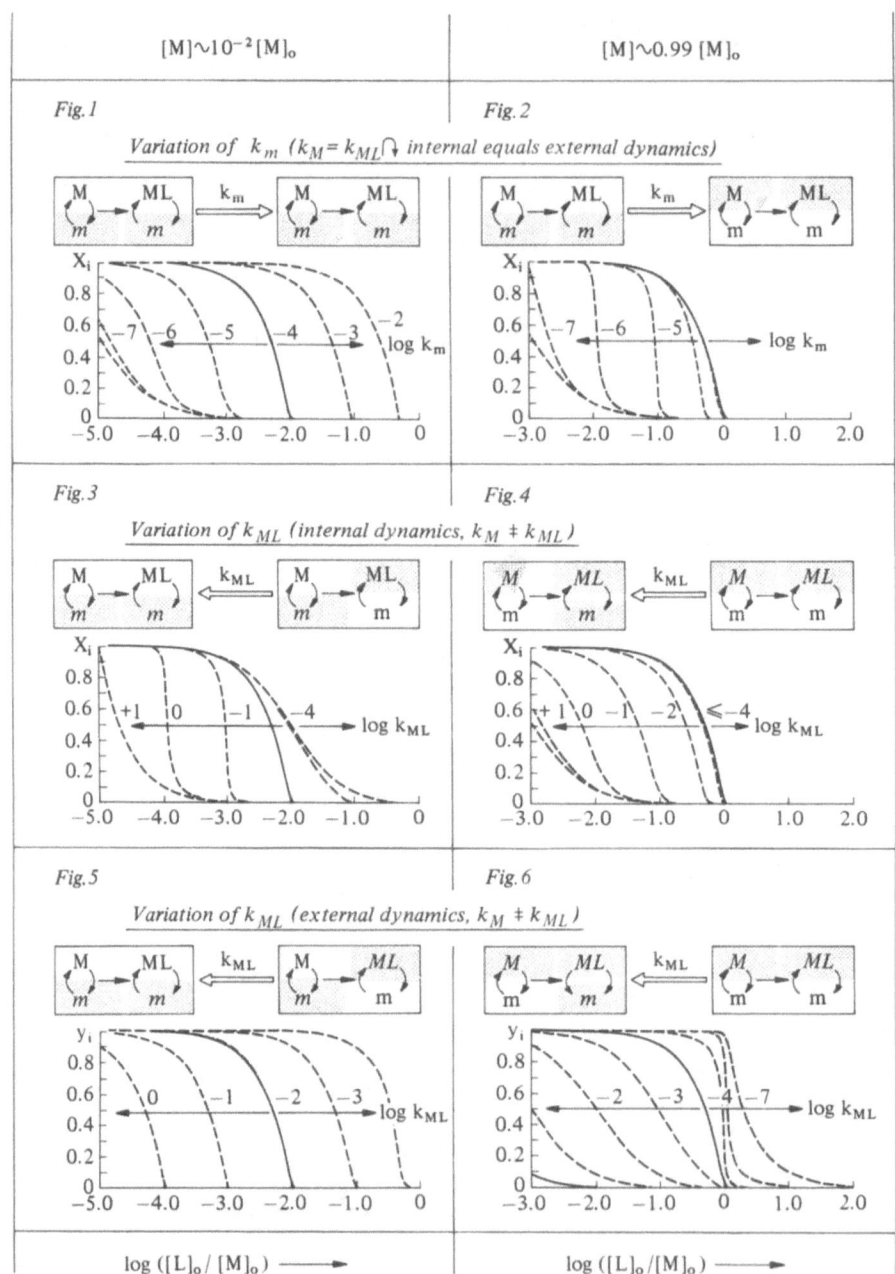

Fig. 1

Fig. 2

Variation of k_m ($k_M = k_{ML}$ ↷ internal equals external dynamics)

Fig. 3

Fig. 4

Variation of k_{ML} (internal dynamics, $k_M \neq k_{ML}$)

Fig. 5

Fig. 6

Variation of k_{ML} (external dynamics, $k_M \neq k_{ML}$)

3.4 Rules for the Interpretation of the Product Pattern in [L]-Control Maps

To obtain information on the "steady-state pattern" of the intermediate complexes one has to analyze position, shape and coupling of the product curves. The following

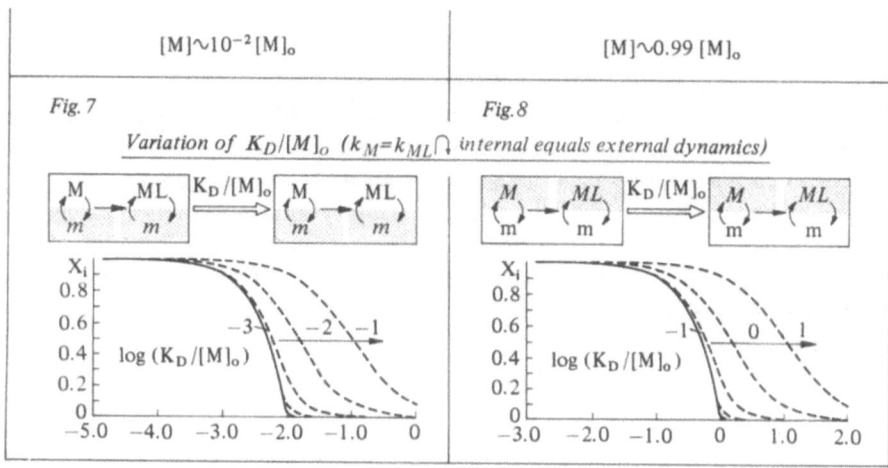

Catalytic model system:

Standard curves (solid lines)

	$[M]\sim 10^{-2}[M]_o$	$[M]\sim 0.99\,[M]_o$
$K_D/[M]_o$	10^{-5}	10^{-3}
k_m	10^{-4}	10^{-2}
k_M	10^{-2}	10^{-4}
k_{ML}	10^{-2}	10^{-4}

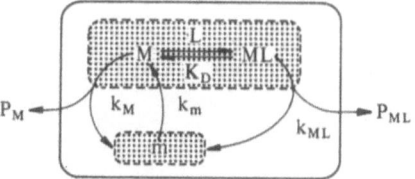

Scheme 3.4–1. Simulated titration curves for the catalytic model system described above. The change in the steady-state concentrations following the ligand association process is schematically depicted (the species present at relatively high concentration is underlined).
Solid lines: standard titration curves, broken lines: manifold systematic variations, arrows: direction of the induced relative shift. Figs. 1 and 2 simulate structural changes in the ligand-free complexes, Figs. 3–6 inhibition and activation processes induced by the controlling ligand (kinetic control), Figs. 7 and 8 simulate a variation of the catalytic concentration (see Scheme 3.3–4) or of the constants of association of L to M (thermodynamic control).

model system of a single product-determining ligand-association process permits to discuss some of the elementary principles for interpreting [L]-control maps.

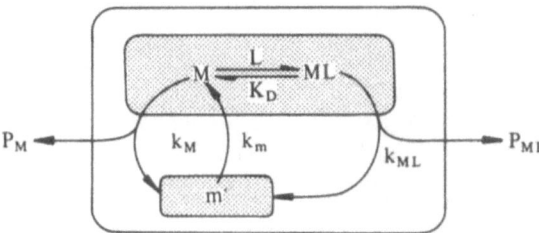

By varying the k values of the kinetically controlled reaction steps in this catalytic model system one can simulate the steady-state situations as well as an activator/inhibitor character of the controlling ligand. There is no distinction between internal and external dynamics of the system if k_M equals k_{ML} (Scheme 3.4-1). If ML is at low steady-state concentration, the titration curves shift to negative $\log([L]_0/[M]_0)$ values. Scheme 3.4-1 demonstrates that the differences between internal and external dynamics (the actual concentration of the species and the resulting selectivity) depend on the order of magnitude of the activation or inhibition process induced by the directing ligand. Even the shape of the titration curve alters characteristically with changing parameters. Scheme 3.4-2 illustrates two different shapes of product curves form [L]-control maps. As can be seen, the shapes of the experimental curves are indeed somewhat related to those of the simulated curves.

Fig. 1. System Ni(COD)$_2$ / pyridine / butadiene (1:x:200)[98]

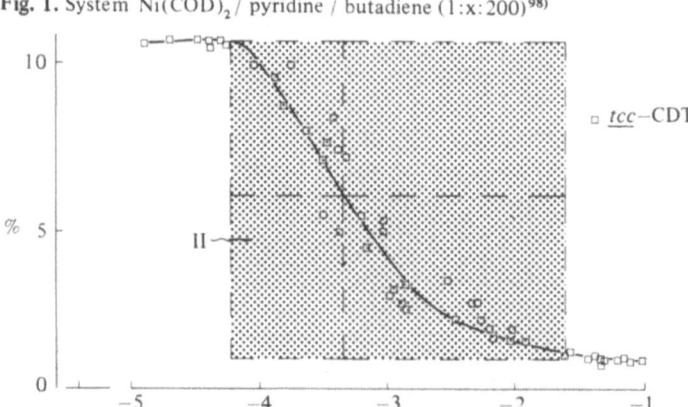

Fig. 2. System Ni(COD)$_2$ / triphenylphosphine / butadiene (1:x:170)[102]

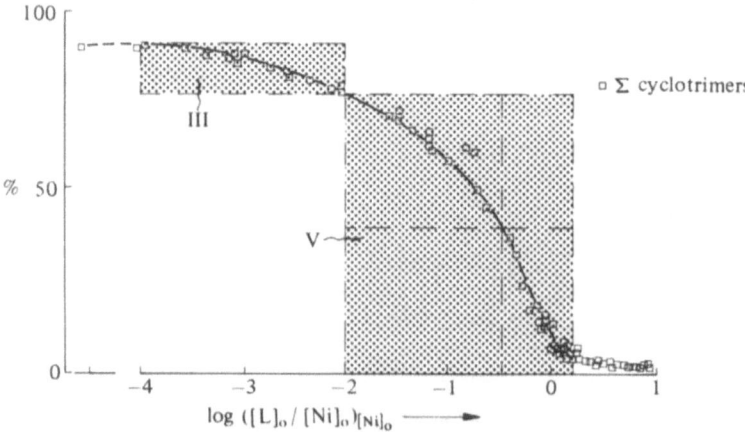

$\log([L]_0/[Ni]_0)_{[Ni]_0}$ ⟶

Scheme 3.4–2. Two different shapes of product-distribution curves from [L]-control maps (product distribution in mol%, reaction conditions see Fig. 3.2–2)

To obtain information on the coupling of the various intermediates one has to analyze the relationship between the corresponding titration curves. Scheme 3.4-3 shows typical steady-state curves for the (1) stepwise twofold association of ligand L with metal complex M, (2) association of L with two metal complexes M and N at equilibrium and (3) association of L to two metal complexes M and N being not at equilibrium (kinetically separated). From these three types of coupling most of the partial maps can be easily interpreted.

Fig. 1

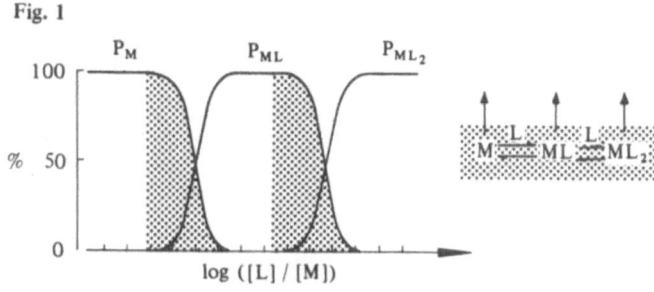

Fig. 2

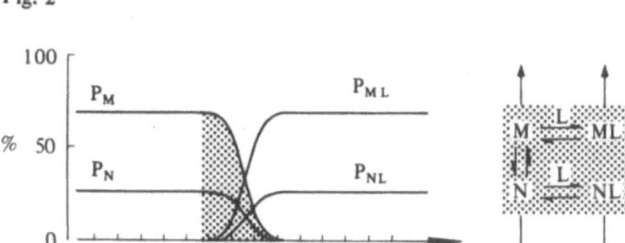

Fig. 3

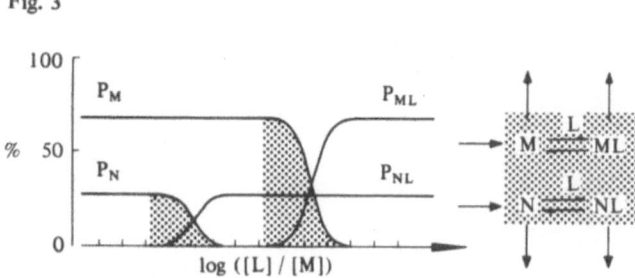

Scheme 3.4–3. Typical selectivity curves of the "titration of a catalytical process" for the stepwise twofold association of ligand L to the intermediate M (Fig. 1), the association of L to two intermediates M and N at equilibrium (Fig. 2) and the association of L to two intermediates M and N being not at equilibrium (Fig. 3) (kinetic separation)

3.5 Ligand-Property Control with Respect to [L]-Control Maps

In the discussion of the property-specific control of a directing ligand in homogeneous transition-metal catalysis, one has to be sure that the results of the considered experiments are comparable. The ligand influences different intermediates of a catalytic system in more or less different ways. This can result in a somewhat contradic-

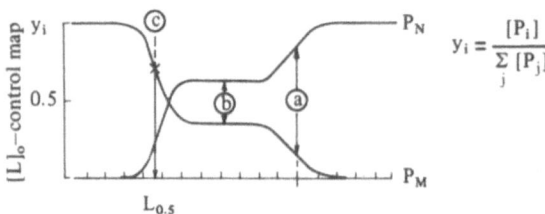

$$y_i = \frac{[P_i]}{\sum\limits_j [P_j]}$$

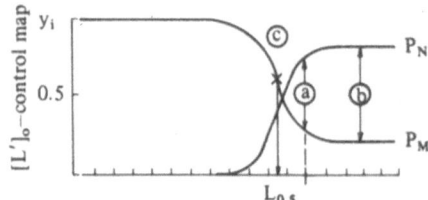

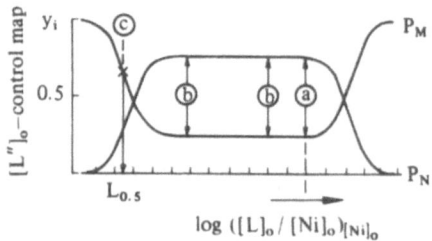

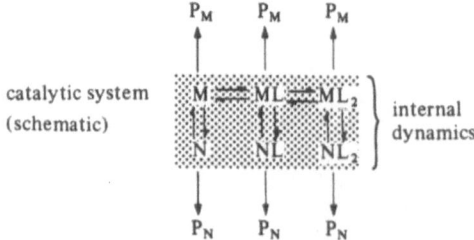

Scheme 3.5−1. Different types of analysis of the property-specific ligand control comparing [L]-control maps for three ligands.

ⓐ Selectivity at constant $\log ([L]_0/[Ni]_0)[Ni]_0$ ratio

ⓑ Selectivity comparing intermediates of the same degree of ligand association

ⓒ Ligand-property induced relative shift of the titration curve for the first association process

tory overall product distribution. On the other hand, the variation of the ligand property can lead to an insight into the in situ-chemical behaviour of individual intermediates if comparable catalytic results are selected.

Three different types of analysis of the property-specific ligand control are schematically described in Scheme 3.5-1 by the [L]-control maps of ligands with quite different association behaviour.

Type (a)
$$y\,(x_j, \Theta_j)[L_j]_0/[Ni]_0 = \text{const.} = \sum_{i=0}^{m} a_i x_j^i + \sum_{i=1}^{n} b_i \Theta_j^i$$

Comparing the product distribution for different ligands at constant $([L]_0/[Ni]_0)[Ni]_0)$ ratio denoted in all three maps by symbol (a), one has to take into account that the resulting selectivity can originate in different types of intermediates. For the given $[L]_0/[Ni]_0$ ratio, the upper schematic [L]-control map reflects the influence of the ligand in the second association step. In the middle one, this ratio is within the range of the first association step. In the map seen below, the resulting selectivity reflects the competition of the mono-ligand associates with one another. Without this information from [L]-control maps, an interpretation of the findings for different ligands at constant $[L]_0/[M]_0$ ratio may be without any predictable power.

Type (b)
$$y\,(x_j, \Theta_j)_{NiLjn}\,(n = \text{const.}) = \sum_{i=0}^{m} a_i x_j^i + \sum_{i=1}^{n} b_i \Theta_j^i$$

If a comparison of the chemical behaviour for intermediates of the same degree of ligand association is desired, the accompanying [L]-control maps must be investigated in order to elucidate the range of existence of the comparable intermediate complexes. Within this range, the kinetic selectivity alone governs the product distribution leading to invariable selectivity. This type of analysis is schematically described in the three maps denoted by (b).

Type (c)
$$\{L_g\,([L_j]_0/[Ni]_0)[Ni]_0\}_{L_{0.5}} = \sum_{i=0}^{m} a_i x_j^i + \sum_{i=1}^{n} b_i \Theta_j^i$$

In [L]-control maps the substitution of one ligand by another one results in a change of the range of existence of the manifold intermediates. This change can be expressed by the ligand-property induced shift of the titration curves identified by the relative position of their inflection points $L_{0.5}$ on the log $([L]_0/[Ni]_0)$ scale. These characteristic shifts provide information on the thermodynamic selectivity governed by the association processes only. This type of analysis is designated by (c).

Following the strategy of analyzing the [L] control by constructing partial maps (see Sect. 3.2), we succeeded in obtaining an insight into the hierarchical order of controlling factors from the view of ligand-property control. This will be demonstrated for the three-component system nickel/P-ligand/butadiene by a type (a) analysis. The directing ligand properties were quantified using Tolman's electronic (χ) and steric (Θ) parameters[62]. If necessary, these effects were separated by a multilinear regression analysis[52, 63].

The property-specific control of the trimer to oligomer ratio is described by the profile of Fig. 1 in Scheme 3.5-2[113]. 11 additional [L]-control maps over the whole range of the ligand properties enable a comparison of the ligand associations at this one-to-one ratio[114].

As is seen this procedure is governed mainly by steric effects of the directing ligands. In contrast, the composition of the cyclodimers is governed only by electronic factors at constant $[L]_0/[Ni]_0$ ratio leading to a U-shaped curve (Fig. 2 in Scheme 3.5-2)[115]. An investigation of the composition of the cyclodimers after the controlling first ligand association (analysis type (b)) results in an invariable COD/VCH ratio of

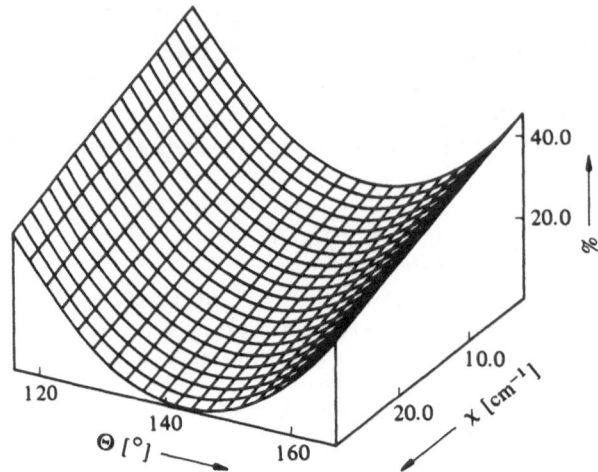

Fig. 1. Profile of the CDT fraction in mol % (y_i=[2+3]/[2+3+5+6])[113]
Variance of fit $S_{1.2}^2$ =30, of measured data δ^2=25 (estimated)
y_i (calc.)=946.8−0.67 x_i −12.968 Θ_i + 453.8 10^{-4} Θ_i^2
(electronic−to−steric ratio = 25 : 75)

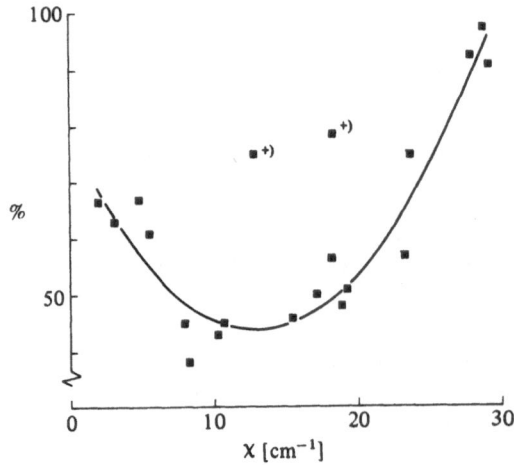

Fig. 2. COD fraction in mol % (y_i = [5]/[5+6])[115]
Variance of fit $S_{2.0}^2$=30, of measured data δ^2=16 (estimated)
y_i (calc.)=79.3−5.43 x_i + 206.3 10^{-3} x_i^2
(100% electronic control), [+] not included

Scheme 3.5−2. Ligand property-specific control as a function of the electronic (x_i) and steric (Θ_i) ligand parameters[62] for the system COD_2Ni/P-ligands/butadiene (1.0:1.0:170), $[Ni]_0$ = 40 mmol/l[118], T = 60 °C, t = 48 h, solvent: toluene, conversion approx. 98%, 20 ligands

Fig. 1. Cyclodimer map (system $COD_2Ni/$ tri$-n-$butylphosphine/butadiene)[114]

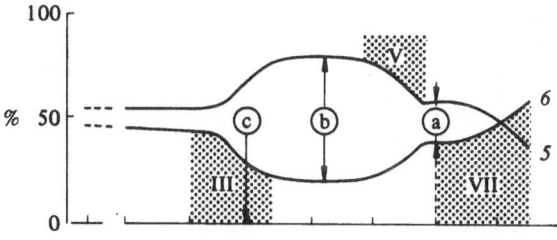

Fig. 2. Cyclodimer map (system $COD_2Ni/$ diphenylethoxiphosphine/butadiene)[114]

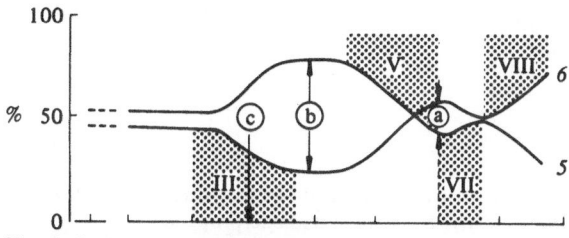

Fig. 3. Cyclodimer map (system $COD_2Ni/$phenyl$-$diphenoxyphosphine/butadiene)[99]

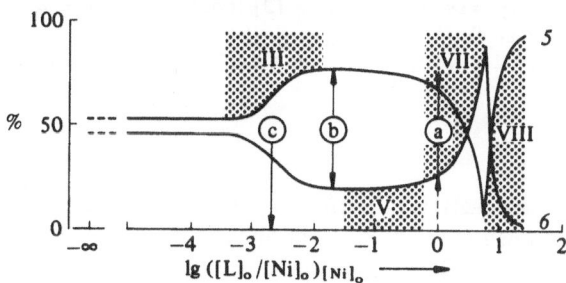

Scheme 3.5–3. Type ⓐ, ⓑ and ⓒ analysis of the property-specific ligand control comparing partial [L]-control maps of the cyclodimer distribution for three different P-ligands (cf. Scheme 3.51–1.) Reaction conditions see Fig. 3.2–2. *5*: COD, *6*: VCH

80/20, independent of the ligand properties (Scheme 3.5-3). Only the position of the corresponding titration curve for this ligand association varies with the added ligands (analysis type ⓒ) shown in Scheme 3.5-3.

Profiles of ligand-property control are as useful for comparing different catalytic systems as ligand-concentration control maps. The ligand control for the amount of dimerization or cyclotrimerization of butadiene and 2:1 co-oligomerization of butadiene and propene is determined in the same subsystem by competition of the P-ligand, butadiene and propene. Therefore, we should obtain closely related profiles between the trimer fraction in the oligomerization of butadiene and the percentage of 2:1 adducts in co-oligomerization. This has been verified (Fig. 1 in Scheme 3.5-2 and 3.5-4).

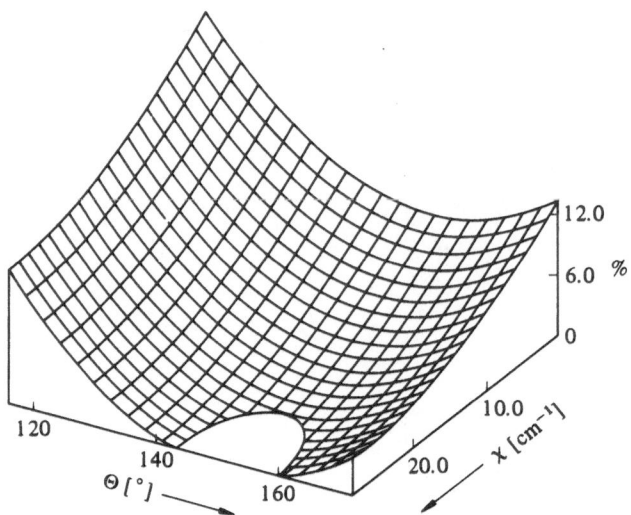

Scheme 3.5–4. Ligand property-specific control as a function of the electronic (x_i) and steric (Θ_i) ligand parameters[62] for the system COD_2Ni/P-ligands/butadiene/propene $(1.0:1.0:200:200)$[103]. $[Ni]_0 = 40$ mmol/l[118], T = 40 °C, t = 144 h, solvent: toluene, butadiene conversion approx. 90%, 17 ligands.
Profile of the co-oligomer fraction in mol% ($y_i = [8 + 9 + 10 + 11 + 12]/[\text{product}]$) variance of fit $S_{2,2}^2 = 5$, of measured data $\sigma^2 = 8$ (estimated)
$y_i = 264,1-68,9 \ 10^{-2} \ x_i + 12,3 \ 10^{-3} \ x_i^2 - 336,9 \ 10^{-2} \ \Theta_i + 111,2 \ 10^{-4} \ \Theta_i^2$ (electronic to steric ratio = 30 : 70)

For complex chemical systems, the analysis of ligand-property control leads to similar profiles. For the above mentioned ligand-concentration control of the system 1,3-dimethylallyl-methyl-nickel/P-ligands/CO, the corresponding profile is shown in Fig. 1 of Scheme 3.5-5. The increase of the cone angle Θ and of the acceptor strength (high χ values) favours C–C bond formation over C=O insertion (15 over 16)[40].

Fig. 1. Profile of the C–C linkage fraction (mol%) ($y_i = [15]/[15 + 16]$) of the reaction products of the system 13/P-ligands/CO $(0.5:1.0:5)$; 18 ligands[40] (reaction conditions see Scheme 3.3–5). Variance of fit $S_{1,2}^2 = 3.6$, of measured data $\sigma^2 = 4$ (estimated)
$y_i = 43,2 + 0,49 \ x_i - 0,743 \ \Theta_i + 33,3 \ 10^{-4} \ \Theta_i^2$ (electronic-to-steric ratio = 55:45)

Fig. 2. Profile of the point of decomposition (Dp) of 14 (0.05 M in toluene) detected by differential thermoanalysis ($\Delta T/\text{min} = +1.5$ °C), 27 ligands[53].
Variance of fit $S_{2,3}^2 = 9$, of measured data $\sigma^2 = 4$ (estimated) (electronic-to-steric ratio 80:20)
$y_i = 110,2 + 1.52 \ x_i - 8,61 \ 10^{-2} \ x_i^2 - 1,29 \ \Theta_i + 1,49 \ 10^{-2} \ \Theta_i^2 - 4,88 \ 10^{-5} \ \Theta_i^3$

Scheme 3.5–5. Ligand property-specific control as a function of the electronic (x_i) and steric (Θ_i) ligand parameter[62] for complex chemical systems ▶

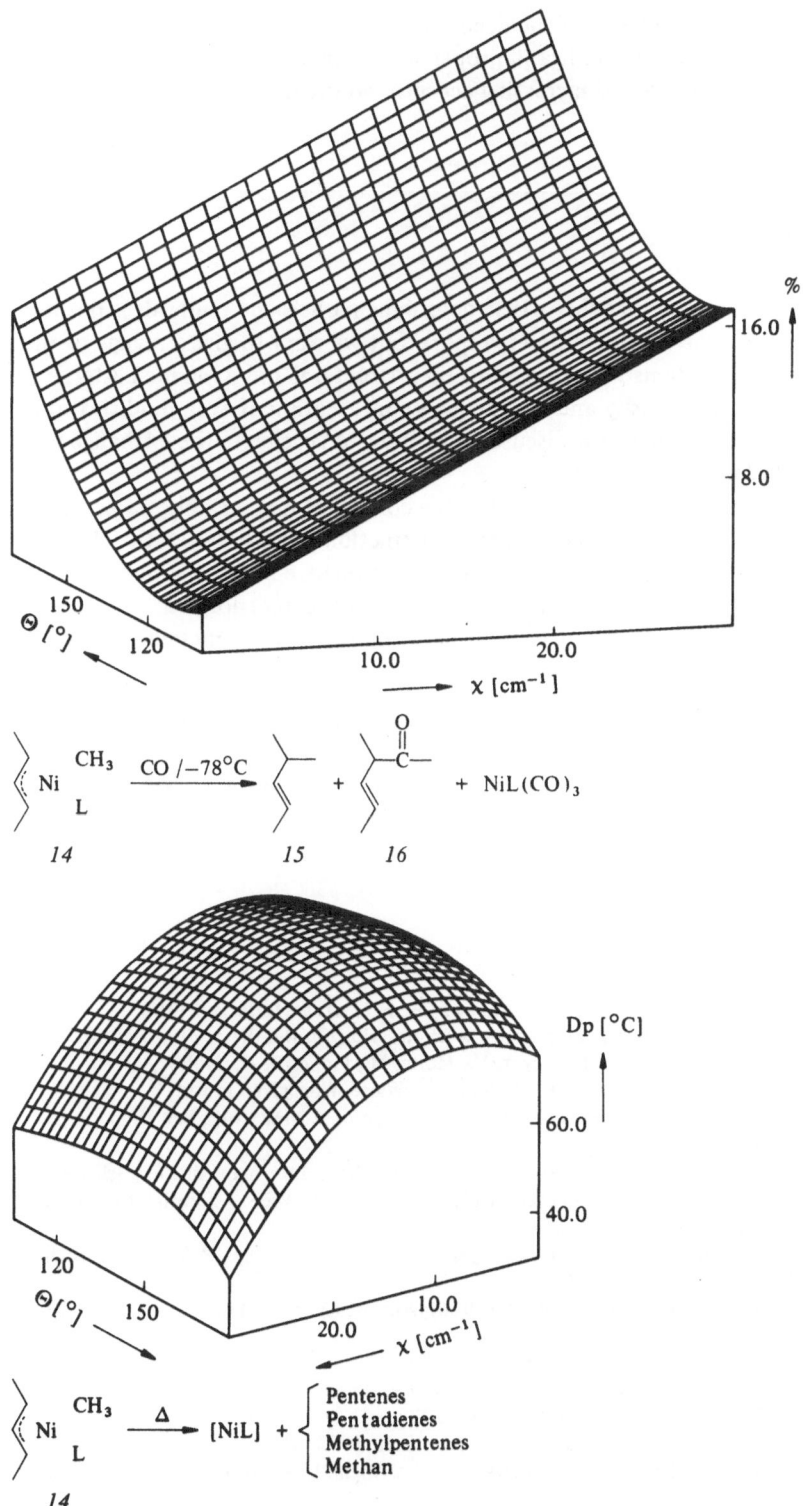

In addition, the dependence of the decomposition temperature (Dp) of the above analyzed 1:1-NiL-complex *14* on ligand properties is shown in Fig. 2 of Scheme 3.5-5. Increasing cone angle and increasing acceptor strength destabilize the tested metal complexes[53].

4 Conclusion

We should like to mention that this article describing models and methods is not intended to give a conclusive and complete report. But we want to give a review of the analyses of complex systems with respect to their structural and dynamic patterns. Some conclusions may be risky and may require confirmation or even modifications. We have raised some problems for discussion in order to accelerate the exchange of information in this fascinating field.

The starting point of our engagement in the control of metal-catalyzed processes was the idea that transition-metal chemistry is a junction point between the classical fields of chemistry (see Scheme 2.1-1). We therefore focussed on the application of the approved rules, models and methods of these areas to metalorganic chemistry. On the other hand, a deeper insight into the controlling factors in metalorganic chemistry will trace back to the origins.

Acknowledgement. We would like to thank all our coworkers for their valuable assistance and Dr. J. Kluth also for the didactic elaboration of the schemes used. Financial support of Deutsche Forschungsgemeinschaft, Landesamt für Forschung des Landes Nordrhein-Westfalen and Fonds der Chemischen Industrie is gratefully acknowledged.

5 References

1. Reed, H. B. W.: J. Chem. Soc. *1954*, 1931
2. Wilke, G.: Angew. Chem. *75*, 10 (1963); Angew. Chem. Internat. Edit. *2*, 1 (1963)
3. A detailed information about the preparative work described in the theses of our group Buchholz, H., Fleck, W., Hey, H.-J., Meyer, V., Molin, M., Ploner, K.-J., Roloff, A., Schimpf, R., Scholz, K.-H., Selbeck, H., Thömel, F., Wiese, W. is given in Ref.[4]
4. Jolly, P. W., Wilke, G.: The organic chemistry of nickel, Vol. 2 New York, San Francisco, London: Academic Press 1975
5. Heimbach, P.: Angew. Chem. *85*, 1035 (1973); Angew. Chem. Internat. Edit. English *12*, 975 (1973); for syntheses from the catalytically formed products see: J. Synth. Org. Chem. Japan *31*, 299 (1973)
6. Heimbach, P.: Aspects of homogeneous catalysis. Vol. 2, p. 81. In: Ugo, R.(ed.), Dordrecht-Holland, Boston (USA): Reidel 1974
7. Heimbach, P., et al.: Angew. Chem. *88*, 29 (1976); Angew. Chem. Internat. Edit. English *15*, 49 (1976)
8. Thömel, F.: Thesis. Ruhr-Universität Bochum, 1970
9. Scholz, K. H.: Thesis. Ruhr-Universität Bochum 1974
10. (a) Molin, M.: Thesis, Ruhr-Universität Bochum 1972
    (b) Ohno, K., Mitsuyasu, T., Tsuji, J.: Tetrahedron Lett. *1971*, 67
11. Heimbach, P.: Angew. Chem. *80*, 967 (1968); Angew. Chem. Internat. Edit. English 7, 828 (1968)

12. Nabbefeld-Arnold, E. F.: Thesis. Universität Essen – GHS, 1980
13. Heimbach, P. et al.: Angew. Chem. *89*, 261 (1977); Angew. Chem. Internat. Edit. English *16*, 253 (1977)
14. Arfsten, N.: Thesis. Universität Essen 1980
15. Bandmann, H., Heimbach, P., Roloff, A.: J. Chem. Res. (S) *1977*, 261; J. Chem. Res. (M) *1977*, 3056-69
16. This principle was in part discussed by P. Heimbach at the FECHEM-conference in Hameln 1978; for first approaches see Refs. [17, 18]
17. Heimbach, P., Traunmüller, R.: Metall-Olefin-Komplexchemie. Weinheim/Bergstr.: Verlag Chemie 1970
18. Traunmüller, R. et al.: Chem. Phys. Lett. *3*, 300 (1969)
19. This method and its application was presented by Schenkluhn, H. (see Ref. [20]) at the FECHEM-conference in Hameln, 1978
20. The development of this method, including simulations and regression analyses, is part of the habilitation of H. Schenkluhn
21. Winkler-Oswatitsch, R., Eigen, M.: Angew. Chem. *91*, 20 (1979); Angew. Chem. Internat. Edit. *18*, 20 (1979)
22. Heimbach, P., Kluth, J., Schenkluhn, H.: Catalysis in chemistry and biochemistry, theory and experiment. In: B. Pullman (ed.), Vol. 12, p. 227. Dordrecht, Boston, London: Reidel 1979
23. We originally called it metalogy principle. However, Jörgensen (Geneve), proposed the term "metala-logy principle", which is more correct and fits better to our expression "metala-ring closure reactions": see Ref. [17]
24. Fukui, K.: Theory of orientation stereoselection, Berlin, Heidelberg, New York: Springer 1975
25. (a) Rösch, N., Hoffmann, R.: Inorg. Chem. *13*, 2656 (1974)
    (b) Elian, M. et al.: Inorg. Chem. *15*, 1148 (1976)
    (c) Thorn, D. L., Hoffmann, R.: Inorg. Chem. *17*, 126 (1978)
26. (a) Thorn, D. L., Hoffmann, R.: J. Amer. Chem. Soc. *100*, 2079 (1978)
    (b) Hoffmann, R., Albright, Th. A., Thorn, D. L.: Pure & Appl. Chem. *50*, 1 (1978)
    (c) Berke, H., Hoffmann, R.: J. Am. Chem. Soc. *100*, 7224 (1978)
27. Epiotis, N. D. et al.: Structural theory of organic chemistry. Berlin, Heidelberg, New York: Springer 1977
28. (a) Heilbronner, E., Bock, H.: Das HMO-Modell und seine Anwendung, 2. Aufl., Weinheim/Bergstr.: Verlag Chemie, 1978
    (b) Bock, H.: Angew. Chem. *89*, 631 (1977); Angew. Chem. Internat. Edit. English *17*, 613 (1977)
29. Dewar, M. S., Dougherty, R. C.: The PMO-theory of organic chemistry. New York, London: Plenum Press 1975
30. Fuson, R. C.: Chem. Rev. *16*, 1 (1935)
31. (a) Bischof, P., Gleiter, R., Haider, R.: Angew. Chem. *89*, 122 (1977); Angew. Chem. Internat. Edit. English *16*, 110 (1977)
    (b) Dürr, H., Gleiter, R.: Angew. Chem. *90*, 591 (1978); Angew. Chem. Internat. Edit. *17*, 559 (1978)
32. Eigen, M., Winkler, R.: Das Spiel, p. 334 ff. München, Zürich: Piper 1975
33. Halevi, E. A.: Angew. Chem. *88*, 664 (1976); Angew. Chem. Internat. Edit. English *15*, 593 (1976); here local S/A or σ/C_2 for structures and processes are explained in detail
34. Gleiter, R.: Angew. Chem. *86*, 770 (1974); Angew. Chem. Internat. Edit. English *13*, 696 (1974)
35. Houk, K. N.: J. Amer. Chem. Soc. *95*, 4092 (1973); Acc. Chem. Res. *8*, 361 (1975)
36. Heimbach, P., Roloff, A., Schenkluhn, H.: Angew. Chem. *89*, 260 (1977); Angew. Chem. Internat. Edit. English *16*, 252 (1977)
37. Albright, Th. A., Hofmann, P., Hoffmann, R.: J. Amer. Chem. Soc. *99*, 7546 (1977)
38. Epiotis, N. D.: Theory of organic reactions. Berlin, Heidelberg, New York: Springer 1978
39. Gerding, B.: Thesis. Universität Essen, 1979

40. Berger, R.: Thesis. Universität Essen, 1980
41. Heimbach, P., Piorr, R., Roloff, A.: unpubl. results
42. Selbeck, H.: Thesis. Ruhr-Universität Bochum, 1972
43. (a) Maitlis, P.: The Organic Chemistry of Palladium, Vols. 1 and 2. New York, London: Academic Press 1971
    (b) Tsuji, J.: Advan. Organomet. Chem. *17*, 141 (1979)
44. Keim, W., Chung, H.: J. Org. Chem. *37*, 947 (1972)
45. Chepaikin, E. G., Khidekel, M. L.: Bull. Akad. Sci. USSR, Div. Chem. Sci. (Translation of Izv. Akad. Nauk USSR, Ser. Khim.) *1972*, 1052
46. Barnett, B. et al.: Tetrahedron Lett. *1972*, 1457
47. Barker, G. K. et al.: J. Amer. Chem. Soc. *98*, 3373 (1976)
48. Heimbach, P., Tani, K., Scheidt, W.: Abstracts of the XXIVth Symposium on Organomet. Chem., Kyoto, Japan, p. 112, 1976
49. Jörgensen, W. L., Salem, L.: The organic chemist's book of orbitals, New York: Academic Press 1973
50. (a) Seebach, D., Enders, D.: Angew. Chem. *87*, 1 (1975); Angew. Chem. Internat. Edit. English *14*, 15 (1975)
    (b) Seebach, D.: Angew. Chem. *91*, 259 (1979); Angew. Chem. Internat. Edit English *18*, 239 (1979)
51. Gerding, B., Heimbach, P., Krüger, C.: unpubl. results
52. Schenkluhn, H. et al.: Angew. Chem. *91*, 429 (1979); Angew. Chem. Internat. Edit. English *18*, 401 (1979)
53. Berger, R., Schenkluhn, H., Zähres, M.: unpubl. results
54. Bogdanovic, B.: Advan. Organomet. Chem. *17*, 105 (1979)
55. Pruett, R. L., Smith, J. A.: J. Org. Chem. *34*, 327 (1969)
56. (a) US P. 3 496 215 (1970), Dupont, USA (inventors: Drinkard, W. C., Lindsey, R. V.)
    (b) DOS 2 237 703 (1973), Dupont, USA (inventors: King, C. M., Seidel, W. C., Tolman, C. A.)
57. Teyssié, Ph.: lecture at the FECHEM-conference in Hameln 1978
58. Pino, P., Consiglio, G.: Fundamental research in homogeneous catalysis, p. 147 ff., New York, London: Plenum Press 1977
59. Strohmeier, W.: Katalysatoren, Tenside und Mineralöladditive, p. 96 ff., Stuttgart: Thieme 1978
60. Tolman, C. A.: J. Amer. Chem. Soc. *92*, 2953 (1970)
61. Tolman, C. A.: J. Amer. Chem. Soc. *92*, 2956 (1970)
62. Tolman, C. A.: Chem. Reviews *77*, 313 (1977)
63. Berger, R., Schenkluhn, H., Weimann, B.: unpubl. results
64. Johannson, G., Stelzer, O., Unger, E.: Chem. Ber. *108*, 1246 (1975)
65. Barbeau, C., Turcotte, J.: Can. J. Chem. *54*, 1603 (1976)
66. Vastag, S., Heil, B., Marko, L.: J. Mol. Cat. *5*, 189 (1979)
67. Ittel, S. D.: Inorg. Chem. *16*, 2589 (1977)
68. Thorsteinson, E. M., Basolo, F.: J. Amer. Chem. Soc. *88*, 3929 (1966)
69. Bodner, G. M.: Inorg. Chem. *14*, 1932 (1975)
70. Mc Daniel, D. H., Brown, H. C.: J. Org. Chem. *23*, 420 (1958)
71. Levitt, L. S., Widing, H. F.: Progr. Phys. Org. Chem. *12*, 122 (1976)
72. Barlin, G. B., Perrin, D. D.: Quart. Rev. *20*, 75 (1966)
73. Kabachnik, M. I., Balueva, G. A.: Bull. Acad. Sci. USSR, Div. Chem. Sci. *1962*, 495
74. Chapman, N. B., Shorter, J. (eds.): Advances in linear free energy relationships, p. 225. New York: Plenum Press 1972
74. Koppel, I. J., Palm, V. A.: Advances in linear free energy relationships, In: Chapman, N. B., Shorter, J. (eds.), p. 225., New York: Plenum Press 1972
75. Bogatkov, S. V., Popov, A. F., Litvinenko, L. M.: Reakts. Sposobnst Organich. Soed. *6*, 1011 (1969)
76. Jones, Th. E., Cole, J. R., Nusser, B. J.: Inorg. Chem. *17*, 3680 (1978)
77. Litvinenko, L. M., Popov, A. F., Gelbina, Zh. P.: Dokl. Chem. *203*, 229 (1972)
78. Mac Phee, J. A., Panaye, A., Dubois, J.-E.: Tetrahedron *34*, 3553 (1978)

79. Vincent, A. T., Wheatley, P. J.: J. Chem. Soc. Dalton Trans. *1972*, 617
80. Reed, F. J. S., Venanzi, L. M., Bachechi, F., Mura, P., Zambonelli, L.: unpubl. results
81. Bracher, G., Grove, D. M., Venanzi, L. M., Bachechi, F., Mura, P., Zambonelli, L.: unpubl. results
82. Astrup, E. E. et al.: Acta Chim. Scandinavia A 29 *1975*, 827
83. Tamaki, A., Magennis, S. A., Kochi, J. K.: J. Amer. Chem. Soc. *96*, 6140 (1974)
84. Heimbach, P. et al: in preparation
85. for ligand/metal separations see also: Mingos, D. M. P.: J. Chem. Soc. Dalton *1977*, 20, 26, 31
86. Batich, C. D.: J. Am. Chem. Soc. *98*, 7585 (1976)
87. Warren, L. F., Bennett, M. A.: Inorg. Chem. *15*, 3126 (1976)
88. Mc Auliffe, C. A. (ed.): Transition metal complexes of phosphorus, arsenic and antimony ligands. London: Mc Millan 1973
89. Dixon, M., Webb, E. C.: Enzymes. 2nd edit. London: Longman 1971
90. Falbe, J.: Carbon monoxide in organic synthesis, Berlin, Heidelberg, New York: Springer 1970
91. Chien, J. C. W. (ed.): Coordination polymerisation. New York: Academic Press 1975
92. Seel, F.: Grundlagen der analytischen Chemie, 5th edit. Weinheim/Bergstr.: Verlag Chemie 1973
93. Boor, Jr., J.: J. Polym. Sci C *1*, 257 (1963)
94. Su, A. C. S., Collette, J. W.: J. Organomet. Chem. *90*, 227 (1975)
95. Rys, P.: Acc. Chem. Res. *9*, 345 (1976)
96. Kucharkowski, R., Drescher, A., Großmann, O.: Z. Chem. *19*, 281 (1979)
97. Using [31] P-NMR spectroscopy we were able to determine the in situ steady-state situation of some Ni-catalytic systems for $[L]_0/[M]_0$ ratios $> 10^{-1}$
98. Brille, F., Kluth, J., Schenkluhn, H.: J. Mol. Cat. *5*, 27 (1979)
99. Kluth, J.: Thesis. Universität Essen, 1980
100. The cyclodimer cis-1,2-divinyl-cyclobutane, which is also formed, rearranges to *5* and *6* during the catalysis and can no longer be detected after the long reaction time chosen (complete conversion of butadiene)
101. We found that many controlling ligands display a selective behavior among the manifold association possibilities. For example, lewis acids like ethyl-ethoxy-acetylacetonatoaluminium fail to undergo associations with bis-π-allyl-Ni-intermediate complexes (II, III); special four-fold chelating ligands make possible exclusively process VI.
102. Brille, F. et al.: Angew. Chem. *91*, 428 (1979), Angew. Chem. Internat. Edit. *18*, 400 (1979)
103. Brille, F., Heimbach, P., Schenkluhn, H.: unpubl. results 1976
104. Fleck, W.: Thesis. Universität Bochum, 1971
105. W. J. Albery and J. R. Knowles [106] introduced the concept of uniform bonds for optimized catalytic systems. The [L]-control maps provide an experimental approach to this concept.
106. Albery, W. J., Knowles, J. R.: Angew. Chem. *89*, 295 (1977), Angew. Chem. Internat. Edit. *16*, 285 (1977)
107. De Haan, R., Dekker, J.: J. Catal. *44*, 15 (1976)
108. (a) Wilke, G. et al.: Angew. Chem. *78*, 157 (1966), Angew. Chem. Internat. Edit. *5*, 151 (1966)
    (b) Karmann, J.: Thesis. Universität Bochum, 1970
109. Heimbach, P., Hübinger, E., Schenkluhn, H.: unpubl. results 1977
110. Sisak, A., Schenkluhn, H., Heimbach, P.: Acta Chim. Acad. Sci. Hung. (in press)
111. Heimbach, P., Schenkluhn, H. et al.: in preparation
112. Grubbs, R. H. et al.: J. Amer. Chem. Soc. *99*, 3863 (1977)
113. Heimbach, P., Kluth, J., Schenkluhn, H.: Angew. Chem. *92* (1980) in press
114. Kluth, J., Schenkluhn, H., Sisak, A.: in preparation
115. Heimbach, P., Kluth, J., Schenkluhn, H.: Angew. Chem. *92* (1980) in press
116. Bogdanović, B.: Liebigs Ann. Chem. *727*, 143 (1969)
117. Büssemeier, B., Jolly, P. W., Wilke, G.: J. Amer. Chem. Soc. *96*, 4726 (1974)
118. $[Ni]_0$ depends on the temperature of the reaction mixture. The given value is for $T = -78\,°C$, the temperature where the standard solutions are mixed ($[Ni]_0^{60\,°C} = [Ni]_0^{-78\,°C}/1{,}18$).

P. Heimbach and H. Schenkluhn

Notes Added in Proof

119. Definitions for LU/LU interactions are given in analogy to definitions of relative DO/ACC character in HO/HO interactions. This type of double dual control is only important in degenerate subsystems! In molecules containing marked DO/ACC differentiations of its subsystems like e.g. arene · Cr (CO)$_3$ dual control is observed[37]. But for (arene)$_2$Cr-complexes with their two arene groups we have to consider more differentiations.
120. Following the concepts on page 53 and 57 we find out a third concept for experimental strategies to change order parameters in the A/S sense regarding structure and reactivity (for definition of A/S see[33]). This concept fits for degenerated subsystems on the level

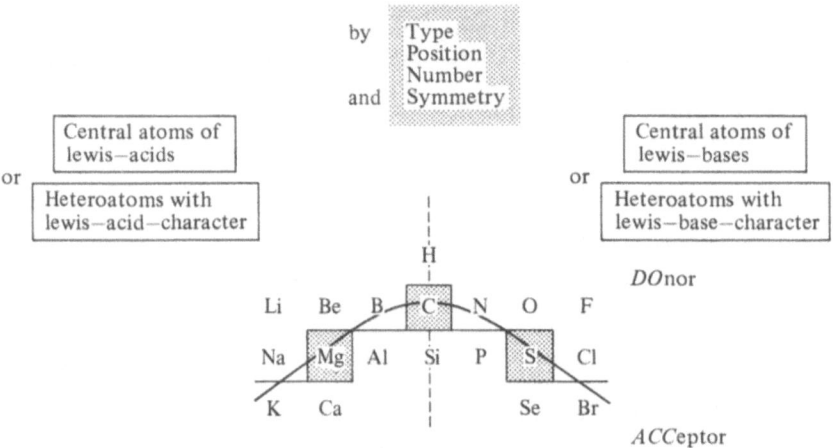

Perturbations:

of molecules (e.g. cis/trans-, staggered/eclipsed- and R/S-differentiations, reactivity of o/p-positions of arenes etc.), cristals (e.g. A/S symmetries in packing effects) and bio-systems (e.g. Na/K- or Ca/Mg-control for structural aspects of biopolymers). The authors of this article assume, that this concept is extremely helpful as a basis for strategies in Umpolung-phenomena of given structures and processes. A documentation is in preparation.

Received January 16, 1980

In Search of New Organometallic Reagents for Organic Synthesis*, [1)]

Thomas Kauffmann

Organisch-Chemisches Institut der Universität Münster, Orléans-Ring 23, D-4400 Münster, Federal Republic of Germany

Table of Contents

---

* This paper was in part presented as a lecture at the meeting "Stereochemical Aspects of Organometallic Reactions" in Hameln/Weser on September 26–29, 1978

1 Introduction

It is a characteristic feature in the development of organic chemistry in the second half of this century that the assortment of functional groups has been enriched by more and more element groups. Normally, as in Scheme 1, they do not appear in the final product of a synthetic operation. In most cases they are introduced in organic compounds for a definite purpose, they work as desired and are eliminated either automatically during the action or in a subsequent special step. The british chemist Warren[2] has nominated those functional groups as "mobile functional groups". The most prominent ones probably are the triphenylphosphonium group (Scheme 1) and the different boranyl groups (Scheme 2). But the opposition of

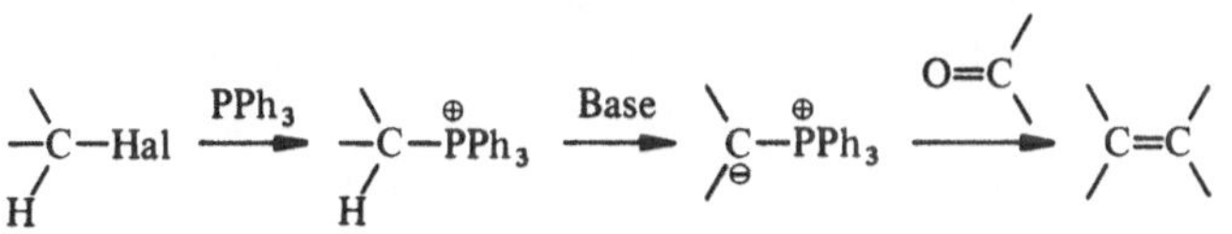

Scheme 1

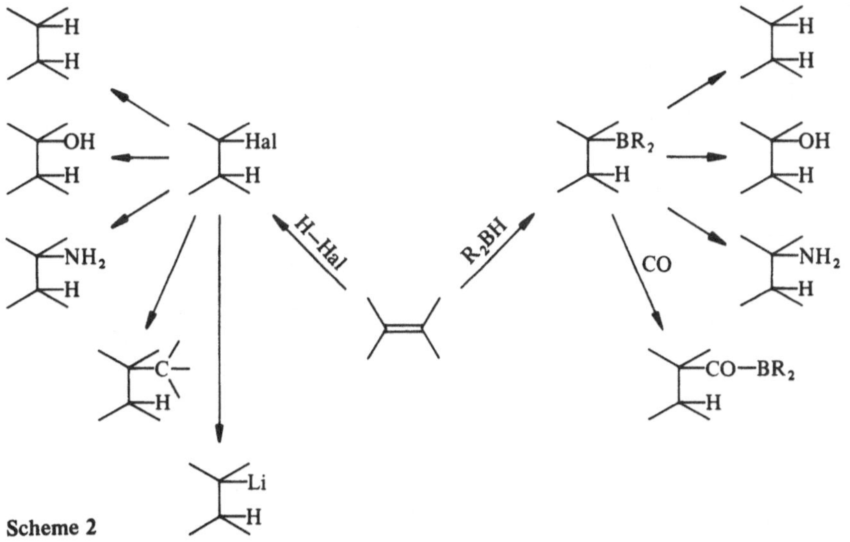

Scheme 2

halogen and boron compounds in Scheme 2 demonstrates clearly that the conception of "mobile functional groups" is by no means very new, because we had similar things with halogen groups for a long time. This treatise attends to new reagents containing organoelement groups. We came across to those reagents during our investigations about 1,3-anionic cycloadditions[3]. The first chapter deals with such reactions.

2 Organoelement Group Assisted 1,3-Anionic Cycloadditions

2.1 Scope of 1,3-Anionic Cycloadditions

The scope of this reaction type, discovered in our laboratory 8 years ago, is strikingly restricted by the fact that aromatic residues are necessary for activating the substrates, for stabilization of the reagents, and in some cases of the products as well.

Cycloadditions of this type do not occur with isolated non-activated C-C double bonds, the 2-azaallyl system (Scheme 3) is only capable of existence with stabilizing aromatic groups[3a)] and the allyl-lithium system (Scheme 3) has no tendency at all to undergo cycloaddition, if a group is lacking which can stabilize the negative charge in the 2-position[4)].

Carbonyl or cyano groups, of proven value in Diels-Alder reactions and 1,3-dipolar cycloadditions, are unsuitable for anionic cycloadditions owing to the pronounced nucleophilic and basic character of the anionic reagents. Instead of cyclo-addition nucleophilic attack of these groups or deprotonation of the substrates would occur. This means that aromatic residues are indispensible which are practically unremovable after the cycloaddition and, unlike the carbonyl group, hardly unsuitable for subsequent synthetic steps.

We asked the question if it is possible to replace the phenyl residues which are parts of the reagents of Scheme 3 by organoelement groups. On these groups we make the following demands: They should be electron withdrawing and resistant to the attack of 2-azaallyl-lithium, of allyl-lithium compounds, of water, and of oxygen as well. Moreover they should be easily removable after the cycloaddition.

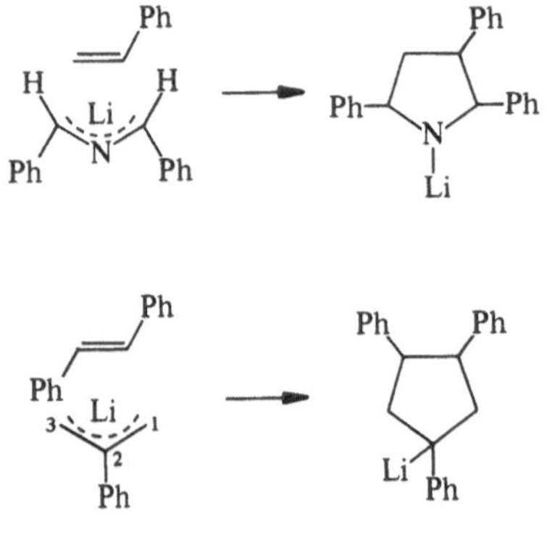

Scheme 3

Therefore organoelement groups with elements of the in Scheme 4 stated area of the periodic table came in question in the first place. The electron withdrawing effect of the uncharged second-row-elements sulfur, phosphorus, and silicium is well-known and mostly ascribed to empty energetic low d-orbitals[5)]. Organoelement groups of the heavier elements were interesting to us because of providing better possibilities for splitting them off after the cycloaddition. Since the electron withdrawing effect of the second-row-elements lately is supposed being due to polariza-

Si	P	S
Ge	As	Se
Sn	Sb	Te
Pb	Bi	

Scheme 4

tion and stereoelectronic effects[6] (see Chap. 5. and Ref.[63]), the third-, forth- and fifth-row-elements should have an analogous effect, as it is already established for selenium[7] and tellurium[8]. It is not very much what we have done and managed in this field. Especially the problem of an universal reagent for the "cyclopentana-tion"[3a] of C-C double bonds is completely unresolved this day. But the setbacks in this field stimulated us to study closer the different organoelement groups and have led to some progress in other fields.

2.2 Organoelement Group Substituted Ethylenes as Substrates

We have found that the groups, scheduled in Table 1, enable a C-C double bond to react with 1,3-diphenyl-2-azaallyl-lithium (*1*) in a cycloaddition reaction[9, 10]. With the exception of the arsenic compound, cycloadducts were formed also with the corresponding 1,1-diphenyl reagent (*2*) which generally has lower tendency to undergo cycloadditions. In this case from both possible regioisomeric cycloadducts was produced only that one with the three phenyl groups at the same side of the ring, indicating that stabilization of negative charge at the carbon atom in α-position to the organoelement group during the cycloaddition process is essential. This is consistent with the two-step mechanism formulated in Scheme 5, but does not rule out a synchronous cycloaddition process which was pointed out by us in other anionic cycloaddition reactions[3].

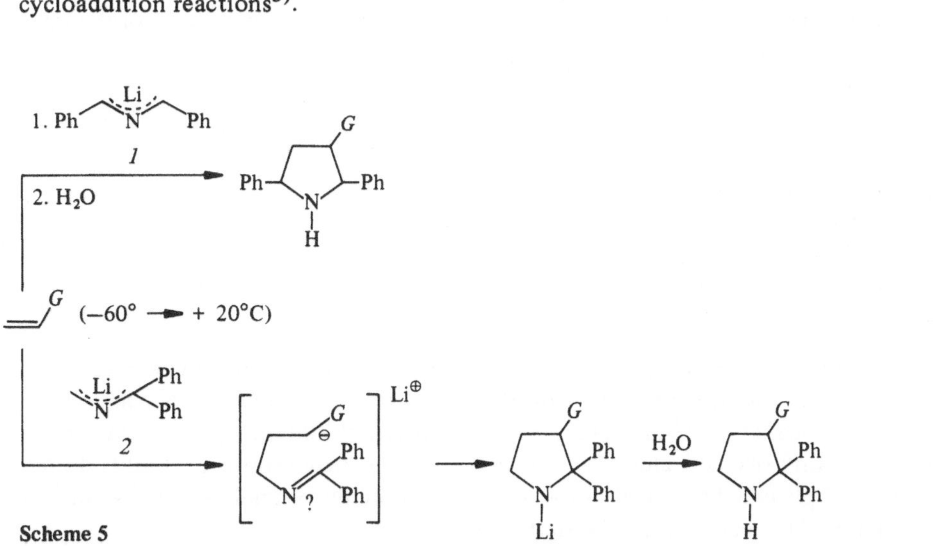

Scheme 5

113

Table 1. Cycloaddition of 2-azaallyl-lithium compounds $CH_2=CH-G$[9,10]

Reagent	G	Yield of cycloadduct (%)
1	−SPh	68
2		64
1	−SePh	77
2		42
1	−PPh$_2$	42
2		65
1	−AsPh$_2$	65
2		0
1	−SiPh$_3$	87[1], 45
2		31
1	−GePh$_3$	29
2		31

[1] Work of E. Popowski[11]

The corresponding organoelement groups of the elements beneath the line in the middle of Scheme 6, for instance the triphenylstannyl group, turned out to be unfit[10a,c]. So we have here a sharp borderline between elements which are suitable in the considered sense and elements which are completely unsuitable. The reason of this is stated in Chap. 6. Converserly to the diphenylphosphanyl group the diphenylphosphoryl group -P(O)Ph$_2$[12] and the analogous group -As(O)Ph$_2$[10d] proved to be unsuitable as activating groups as well. The reason is that C-C double bonds by these groupings are so strongly polarized that anionic polymerization of the vinyl compounds occurs.

Si	P	S
Ge	As	Se
Sn	Sb	Te
Pb		

Scheme 6

In order to get some quantitative information about the activating power of organoelement groups compared with the phenyl residue we did several competition experiments[10b, 13] in which 1,3-diphenyl-azaallyl-lithium (*1*) was reacted in each case with two activated vinyl compounds in the mole ratio 1:1:1 as it is shown in an example in Scheme 7. By those experiments we determined the relative rate constants stated beneath the reaction scheme.

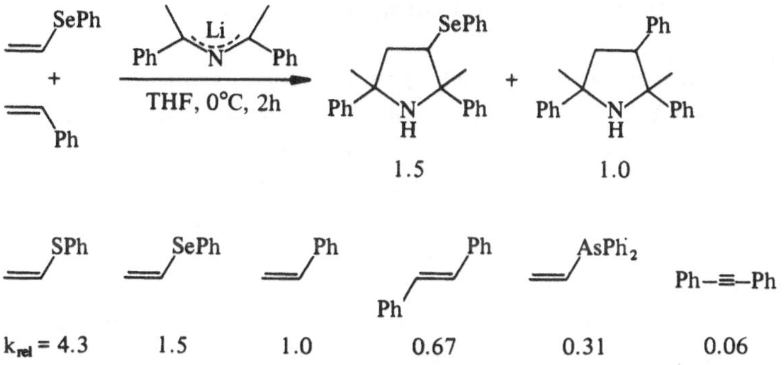

$k_{rel} = 4.3 \qquad 1.5 \qquad 1.0 \qquad 0.67 \qquad 0.31 \qquad 0.06 \qquad$ **Scheme 7**[10a, 10b)]

It became evident that the activation is of the same order of magnitude as with the phenyl residue and that the activating effect decreases distinctly when passing from a second-row-element to a third-row-element.

Organoelement groups like the phenylthio-[14a)] or diphenylarsanyl group[14b)] can be removed reductively from alkyl residues very well by Raney-Nickel (Scheme 8). But the transfer of this splitting reaction to our products of anionic cycloadditions – pyrrolidine derivatives – led to difficulties because, as demonstrated in Scheme 8, in addition to the splitting of the element carbon bond ring opening or aromatization occured.

After methylation of the pyrrolidine N-atom the exchange of the diphenylarsanyl group for hydrogen worked well[10d)]. The corresponding reaction with the N-methyl derivative of 2,5-diphenyl-3-phenylthio-pyrrolidine is still under investigation.

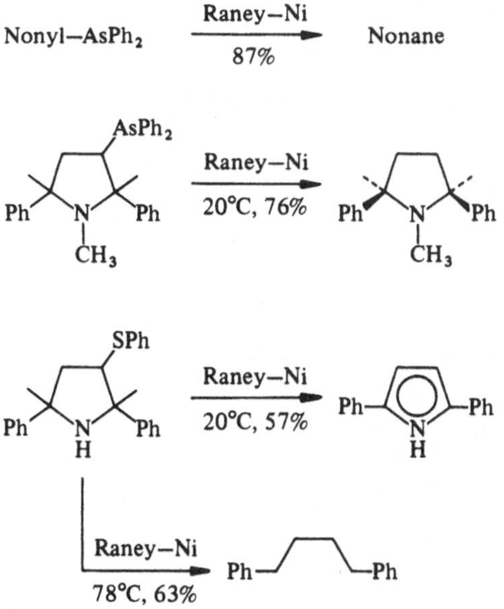

Scheme 8[10b, 10d)]

115

2.3 Organoelement Group Substituted 2-Azaallyl-lithium as Reagents

Fixation of organoelement groups at the 2-azaallyl system deserves some interest too. Today we cannot report about 2-azaallyl systems which have organoelement groups *instead* of the phenyl residues. But the easily accessible lithium compounds 1,1-diphenyl-3-triphenyl-silyl-2-azaallyl-lithium (*4*) and 1,1-diphenyl-3-diphenylphos-phoryl-2-azaallyl-lithium (*5*) proved to be good reagents for transformation of carbonyl compounds to 2-azadienes[15] (Scheme 9). The yields are much better than according to the route *2*⟶ *3*⟶ *6* published by us some years ago[16]. But we are by no means content with this result, because the scope of this synthesis is limited by the two phenyl residues which have been brought in product *6* and analogous products by the reagents *4* and *5*. Efforts to synthesize 2-azaallyl-lithium compounds of the type *7a* and *7b* (G = SiPh$_3$, P(O)Ph$_2$; G' = removable organoelement group) as reagents for syntheses of 2-azadienes were not successful up to now. Many reagents are known in literature which can be applied only in a small range in organic synthesis owing to phenyl residues which are introduced in the product by the reagent. It is to be expected that some of them in the future will be improved by replacing the phenyl residues by removable organoelement groups.

	G	
4	−SiPh$_3$, 58%	
5	−P(O)Ph$_2$, 31%	

Scheme 9 (LDA = lithiumdiisopropylamide)

3 Element/Halogen Exchange in Organic Synthesis

3.1 Search for Halogen Equivalents

Hitherto application of the elements listed in Scheme 4 as organoelement groups in organic synthesis is nearly confined to the light-weight elements Si, P, S, Se. The only exception is tin which has found use in preparative valuable transmetalation reactions[17] and in the synthesis of new heteroaromatic systems as well[18]. Thus mainly the heavier elements were attractive to us. Examination of literature had revealed that organoelement groups of these heavier elements, if combined with a carbon atom, can be replaced not only by hydrogen (see Chap. 2) but also by halogen or lithium. Moreover there are known some β-elimination reactions of organotin and organolead groups together with an hydroxy group resulting an olefine, and last but not least stabilization of a negative charge in α-position is important since allowing deprotonation of an adjacent CH-group or nucleophilic attack to an attached vinyl group. This report is organized by these elementary possibilities summarized in Scheme 10. In this chapter we shall deal with element/halogen exchange in organic

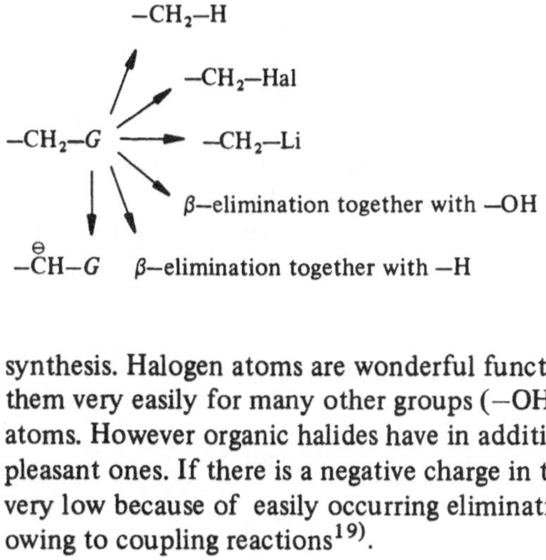

Scheme 10

synthesis. Halogen atoms are wonderful functional groups, because one can replace them very easily for many other groups (−OH, −SH, −NH$_2$, −Li etc.) or carbon atoms. However organic halides have in addition to pleasant properties some unpleasant ones. If there is a negative charge in the compound, stability is normally very low because of easily occurring elimination reactions, α- or β-elimination, or owing to coupling reactions[19].

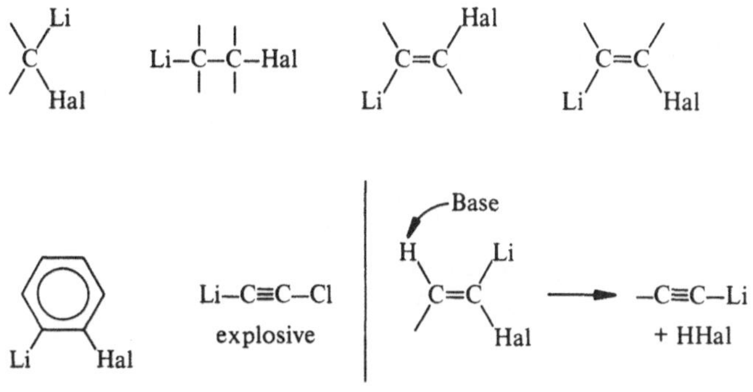

Scheme 11

117

Thus the lithium compounds and groupings put together in Scheme 11 are very unstable and can be used, if at all, only at very low temperature as nucleophilic reagents. The german chemist Köbrich[19] did a very fine work when he nevertheless succeeded in finding out some preparative applications of such compounds.

On account of this instability it is most lucky that various elementorganic groups, for instance the diphenylarsanyl group or phenylselenyl group, can act as halogen equivalents. Such equivalents have to fullfil mainly three suppositions:
1. Introduction on organic compounds should be easy.
2. Stability of anionic compounds with such halogen equivalent groups should be sufficiently high.
3. Exchange of the halogen equivalent group for halogen itself should be easy.

Only the third supposition is critical. Let us therefore look closer at the exchange of elementorganic groups for halogen. The most important case is the exchange of a group attached to a *primary carbon atom* (Scheme 12). The exchange is less prob-

$$\text{n-Hexyl-}G \xrightarrow{\text{Hal}_2} \text{n-Hexyl-Hal} + \text{Hal-}G \qquad\qquad \textbf{Scheme 12}$$

lematic for a group attached to a secondary or tertiary carbon atom. According to our experiments (Table 2) the best yields are provided with the arsenic containing compounds. The diphenylstibanyl group is less favorable because bromolysis needs an essential higher temperature than in case of the diphenylarsanyl group.

Why is the yield so bad in the case of the triphenylstannyl group? The reason is that in this case another cleavage mechanism is operating. If the heteroatom offers a lone pair of electrons, the electrophilic reagent attacks there. An adduct is formed which can be isolated easily in case of organoarsenic[24a], organoantimony[24b], and organoselenium compounds[22]. By heating it suffers fragmention according or analogous to Scheme 13a. These reactions lead exclusively or nearly exclusively to the wished for splitting between the heteroatom and the alkyl residue.

$$\text{Alk-}\overset{\bullet\bullet}{\text{As}}\text{Ph}_2 \xrightarrow{\text{Br}_2} \text{Alk-}\underset{\overset{|}{\text{Br}}}{\overset{\overset{\text{Br}}{|}}{\text{As}}}\text{Ph}_2 \xrightarrow{\Delta T} \text{Alk-Br} + \text{BrAsPh}_2 \qquad\qquad \textbf{Scheme 13a}$$

Table 2. Conditions and yields of bromolysis reactions according Scheme 12

-G	Temp. (°C)	Solvent	Yield of n-Hexyl-Br (%)
-AsPh$_2$	185	No solvent	96[20]
-SbPh$_2$	220	No solvent	65[21]
-SePh	80	EtOH/H$_2$O	44[22]
-SnPh$_3$	23	CCl$_4$	2[23]

If the heteroatom provides no lone pair, as in case of a stannyl[24c)] or plumbyl[24d)] group, bromine and other electrophilic reagents attack preferably the phenyl residues resulting cleavage of the linkage between the heteroatom and the phenyl residue according or analogous to Scheme 13b. If one would take the tributylstannyl group instead of the triphenylstannyl group primary attack of bromine would occur at the Sn-atom analogous to Scheme 13a. But all element carbon bonds would be cleaved afterwards nearly with the same rate resulting once more a low yield of the wanted product.

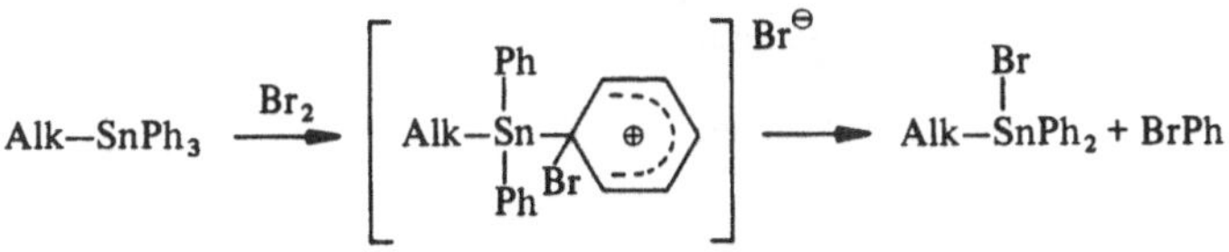

$$\text{Alk-SnPh}_3 \quad \xrightarrow{\text{Br}_2} \quad \left[\text{Alk-Sn} \underset{\substack{| \\ \text{Ph} \\ \text{Br}}}{\overset{\substack{\text{Ph} \\ |}}{}} \right]^{\text{Br}^{\ominus}} \quad \longrightarrow \quad \text{Alk-SnPh}_2 + \text{BrPh}$$

Scheme 13b

Such cleavage reactions at organoarsenic compounds have been known for many years[24a)], but hitherto were only used to prepare the inorganic fragments and not the organic ones. The same holds true for Sb, Sn, Pb or Si containing organoelement groups[24b-e)].

Favorable elements for the element/halogen exchange at alkyl groups are according to our investigations localized in the marked small area of Scheme 14.

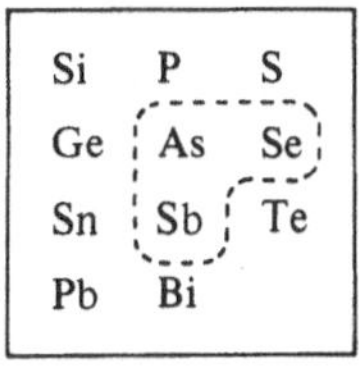

Si	P	S
Ge	As	Se
Sn	Sb	Te
Pb	Bi	

Scheme 14

3.2 Reagents for Nucleophilic Haloalkylation

a) Lithiomethyl-diphenylarsaneoxide *10*

We now turn to reagents which are important in context with the element/halogen exchange: Compounds of the type *8* would be interesting reagents for *nucleophilic halomethylation* but are quite unstable due to carbene formation. Compound *9* would be an attractive synthetic equivalent for the synthons *8*. As will be shown

Hal–CH$_2$–Li   Ph$_2$As–CH$_2$–Li

8   *9*

in Chap. 4 we succeeded in preparing and applying this reagent. The corresponding oxide however, compound *10,* is a much better accessible today and is more reactive. It is formed quantitatively by reacting diphenyl-methyl-arsaneoxide, easily prepared

119

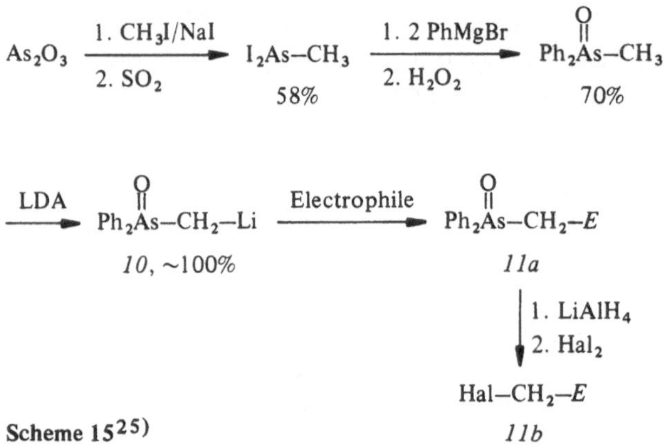

T. Kauffmann

$$As_2O_3 \xrightarrow[\text{2. SO}_2]{\text{1. CH}_3\text{I/NaI}} I_2As{-}CH_3 \xrightarrow[\text{2. H}_2O_2]{\text{1. 2 PhMgBr}} Ph_2\overset{\overset{\displaystyle O}{\|}}{As}{-}CH_3$$

$$\phantom{As_2O_3 \xrightarrow[\text{2. SO}_2]{\text{1. CH}_3\text{I/NaI}} I_2As{-}CH_3}58\% 70\%$$

$$\xrightarrow{\text{LDA}} Ph_2\overset{\overset{\displaystyle O}{\|}}{As}{-}CH_2{-}Li \xrightarrow{\text{Electrophile}} Ph_2\overset{\overset{\displaystyle O}{\|}}{As}{-}CH_2{-}E$$

$$10,~\sim100\% 11a$$

$$\Big\downarrow \begin{array}{l}\text{1. LiAlH}_4 \\ \text{2. Hal}_2\end{array}$$

$$\text{Hal}{-}CH_2{-}E$$

Scheme 15[25)] $11b$

as shown in Scheme 15 with lithiumdiisopropylamide[10f, 25)]. Reacting of *10* with electrophiles and subsequently with lithiumalanate and halogen (Br_2, I_2) gives halogen compounds in preparatively useful yields. If residue E contains a hydroxy group, halogenolysis in some cases gives rise to olefine formation owing to dehydration (Scheme 16[25)]). The yield determining step in reactions *10* $\longrightarrow$ *11a* $\longrightarrow$ *11b* is the formation of *11a*. Table 3 informs about the electrophiles brought to reaction with reagent *10* and about the yields (53–81%) of substitution products *11a*.

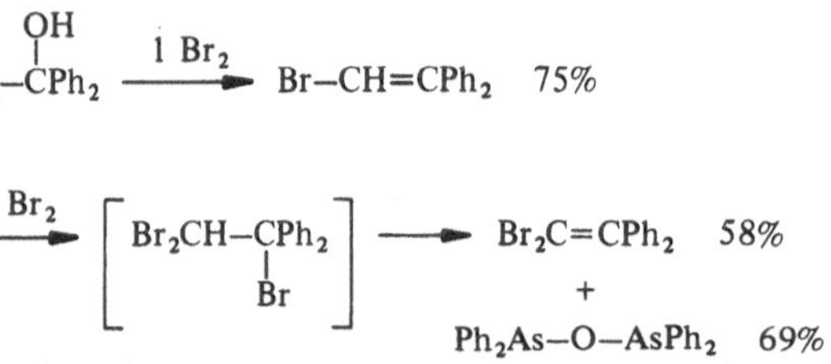

$$Ph_2As{-}CH_2{-}\overset{\overset{\displaystyle OH}{|}}{C}Ph_2 \xrightarrow{\text{1 Br}_2} Br{-}CH{=}CPh_2 \quad 75\%$$

$$\Big\downarrow \text{2 Br}_2$$

$$\left[\begin{array}{c} Br_2CH{-}CPh_2 \\ | \\ Br \end{array} \right] \longrightarrow \begin{array}{l} Br_2C{=}CPh_2 \quad 58\% \\ + \\ Ph_2As{-}O{-}AsPh_2 \quad 69\% \end{array}$$

Scheme 16

Table 3. Reactions of lithiomethyl-diphenyl-arsaneoxide with electrophilic compounds[25)]

Reactant	Product $Ph_2\overset{\overset{\displaystyle O}{\|}}{As}{-}CH_2{-}E$	Yield (%)
n-Butylbromide	$E = -(CH_2)_3{-}CH_3$	72
Allylbromide	$-CH_2{-}CH{=}CH_2$	70
Benzylbromide	$-CH_2{-}Ph$	61
Butyraldehyde	$-CH(OH)(CH_2)_2CH_3$	60
Benzaldehyde	$-CH(OH)Ph$	82
Cyclohexanone	$-C(OH(CH_2)_5$	53
Benzophenone	$-C(OH)Ph_2$	81

b) α-Lithioalkyl-diphenyl-arsaneoxides

By reacting the higher homolges of the methyl-diphenyl-arsaneoxide with lithium-diisopropylamide and then with electrophiles (Scheme 17) compounds of the type *12a* are accessible in similar yields (Table 4[26]).

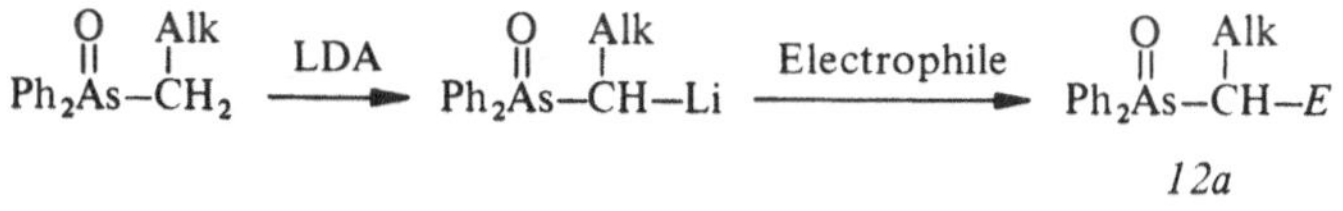

$$
\underset{Ph_2As-CH_2}{\overset{O\ \ \ Alk}{\overset{\|\ \ \ \ |}{}}} \quad \xrightarrow{LDA} \quad \underset{Ph_2As-CH-Li}{\overset{O\ \ \ Alk}{\overset{\|\ \ \ \ |}{}}} \quad \xrightarrow{Electrophile} \quad \underset{Ph_2As-CH-E}{\overset{O\ \ \ Alk}{\overset{\|\ \ \ \ |}{}}}
$$

12a Scheme 17[21, 26]

Table 4. Yields of products 12a[26]

Alk	Electrophile	Yield (%)
C_2H_5	$I-CH_3$	72
C_3H_7	$O=CPh_2$	45
C_4H_9	$Br-(CH_2)_3CH_3$	51
C_4H_9	$O=CPh_2$	93
C_4H_9	CO_2	64

Reduction of compounds *12a* with $LiAlH_4$ ($\sim 100\%$) and following reaction with halogen gave halogen derivatives. Sulfurylchloride was the best reagent for introduction of chlorine, but bromine and iodine gave better yields of the corresponding derivatives. As demonstrated in two examples, one can combine the halogenolysis in a one-pot-reaction with a nucleophilic substitution in order to obtain products of this type. Hydrogen/lithium exchange is even possible at tertiary carbon atoms in α-position to the arsaneoxide group (Scheme 19[26]). So all three hydrogen atoms of reagent *10* can be replaced by electrophilic groups.

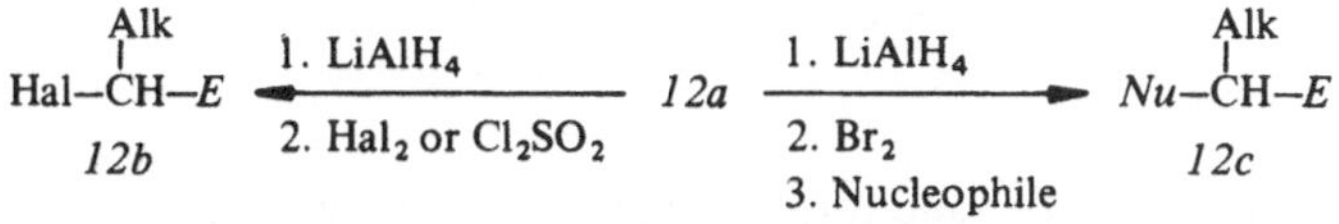

$$
\underset{\underset{12b}{Hal-CH-E}}{\overset{Alk}{\overset{|}{}}} \quad \overset{1.\ LiAlH_4}{\underset{2.\ Hal_2\ or\ Cl_2SO_2}{\xleftarrow{\hspace{2cm}}}} \quad 12a \quad \overset{1.\ LiAlH_4}{\underset{\substack{2.\ Br_2\\3.\ Nucleophile}}{\xrightarrow{\hspace{2cm}}}} \quad \underset{\underset{12c}{Nu-CH-E}}{\overset{Alk}{\overset{|}{}}}
$$

Scheme 18

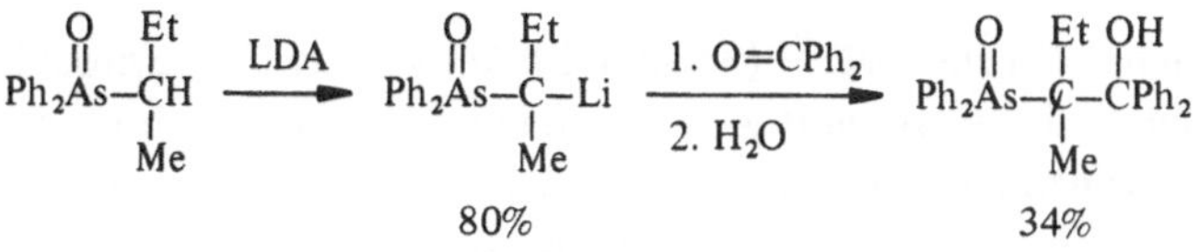

$$
\underset{\underset{Me}{Ph_2As-CH}}{\overset{O\ \ \ Et}{\overset{\|\ \ \ |}{}}} \quad \xrightarrow{LDA} \quad \underset{\underset{Me}{Ph_2As-C-Li}}{\overset{O\ \ \ Et}{\overset{\|\ \ \ |}{}}} \quad \overset{1.\ O=CPh_2}{\underset{2.\ H_2O}{\xrightarrow{\hspace{1.5cm}}}} \quad \underset{\underset{Me}{Ph_2As-C-CPh_2}}{\overset{O\ \ \ Et\ OH}{\overset{\|\ \ \ |\ \ \ |}{}}}
$$

80% 34% Scheme 19

The equivalence *13a* ≡ *13b* (synthons and reagents for nucleophilic haloalkylation) which was demonstrated by our experiments is very similar to the well-known equivalence *13c* ≡ *13d* (synthons and reagents for nucleophilic acylation) shown by

Table 5. Reactions with *12a* according Scheme 18[26)]

12a ⟶ *12b:*

Alk	E	Reagent	Hal	Yield (%)
C_2H_5	CH_3	Cl_2SO_2	Cl	77
C_2H_5	CH_3	Br_2	Br	93
C_2H_5	CH_3	I_2	I	95
C_4H_9	COOH	Br_2	Br	56

12a ⟶ *12c:*

Alk	E	Nucleophile	Nu	Yield (%)
C_2H_5	CH_3	KOH	OH	64
C_2H_5	CH_3	NaSPh	SPh	71

Seebach et al.[27)]. The common features are a negative charge, a strong electrophilic group, and one of the most important functional groups of organic chemistry in both cases. But nucleophilic acylation of course is more important than nucleophilic halo-alkylation. With the Schemes 20–24 will be shown how the conception outlined before can be further developed.

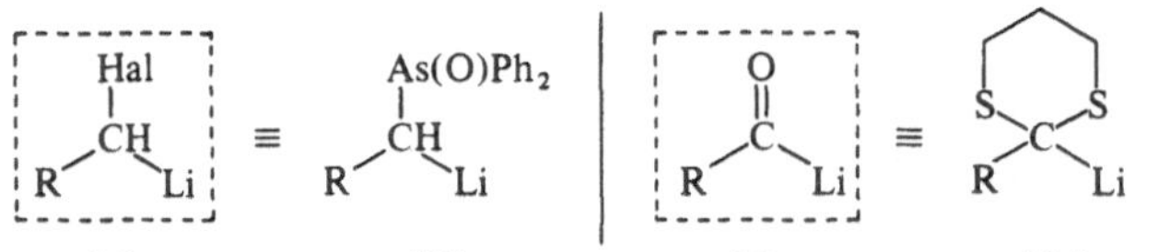

13a *13b* *13c* *13d*

c) α-Lithioallyl-diphenyl-arsaneoxide *14a* and α-Lithiocyclopropyl-diphenyl-arsaneoxide *14b*

To the reagents *14a*[28)] (Scheme 20) and *14b*[28)] (Scheme 21) only short notice should be given since the halogenolysis of the products is not studied up to now. But bromo-lysis of the starting materials which are after reduction to the corresponding arsane comparable to the products, shows that the halogenolysis is possible in case of the allyl group without attack of bromine to the C-C double bond (Scheme 20), whereas in case of the cyclopropyl group bromolysis gives rise to cleavage of the cyclopropane ring resulting a tribromo compound (Scheme 21). With an excess of bromine one can transform the allylarsane to a tribromo compound in good yield too.

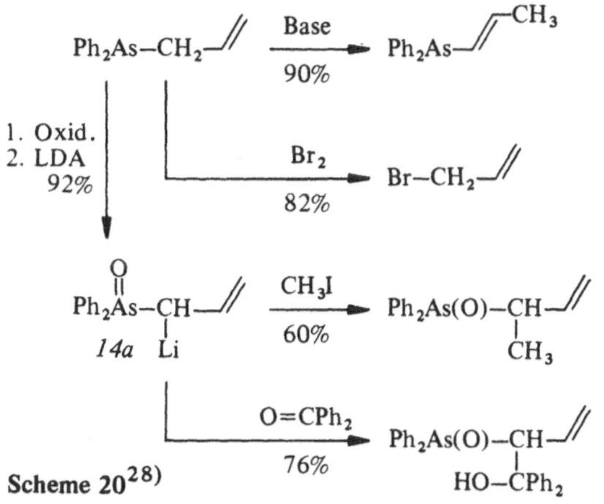

Scheme 20[28)]

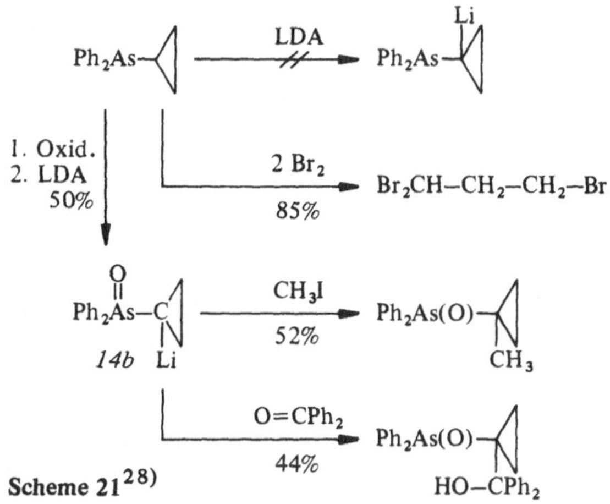

Scheme 21[28)]

3.3 Reagents for Nucleophilic Halovinylation

a) β(Diphenylarsanyl)vinyllithium *16a*

β-Lithiated alkenyl halides of type *15a* would be helpful in organic synthesis. But due to the high instability of these compounds it is not possible to apply them as reagents. We have prepared the synthetic equivalent *16a* by hydrostannation of diphenylarsanylacetylene followed by tin lithium exchange with butyllithium[12)]. Reacting the lithium compound *16a* or the corresponding copper derivative *16b*, quantitatively formed by transmetalation, with electrophiles and subsequently with bromine has yielded the bromides *17a* and *17b*[12)]. So we have a useful ersatz reaction for the nonpracticable reaction *15a* ⟶ *15b*. Reacting with cupricchloride instead of cuprochlorid afforded a diene with two terminal diphenylarsanyl groups, but in this case we could not achieve an arsenic/bromine exchange because polymerization occurred[12)].

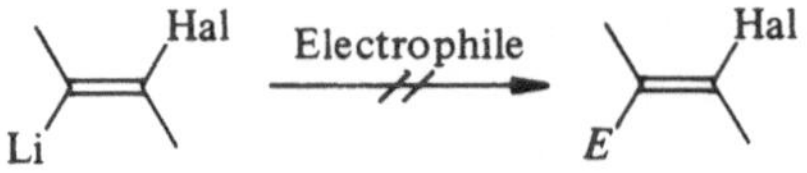

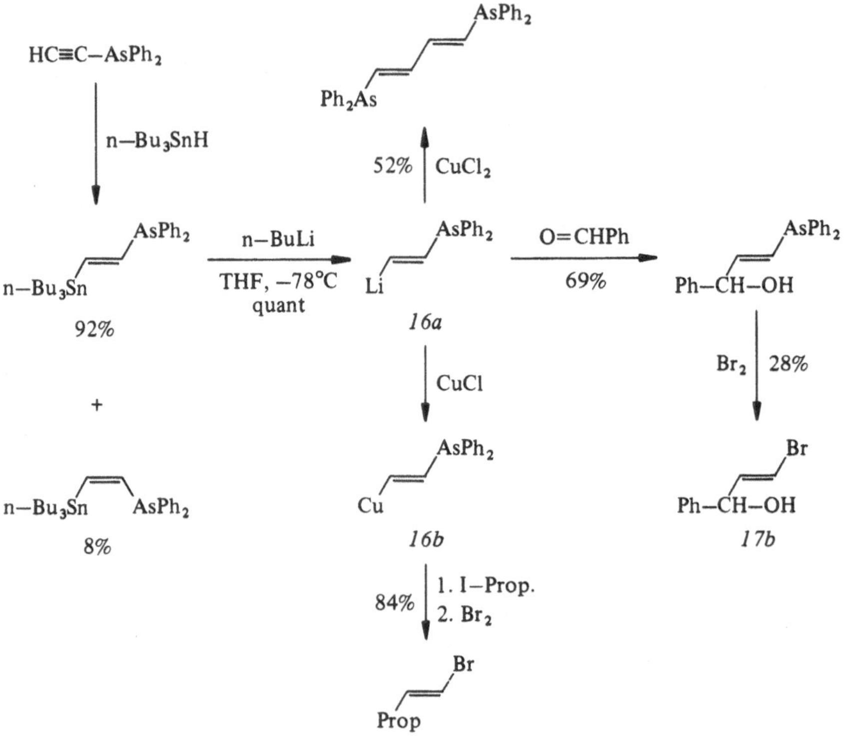

Scheme 22[12)]

b) α(Phenylselenyl) vinyllithium 19a

In search of synthetic equivalents for compounds of type *18a*, which are accessible[19)] but revealed to be stable only at very low temperatures (about −90 °C) we have tried to prepare the lithium compound *18b*, but lithiation of diphenylvinylarsane did not work. In contrast it was possible − as has been found by the group of Krief[29)]

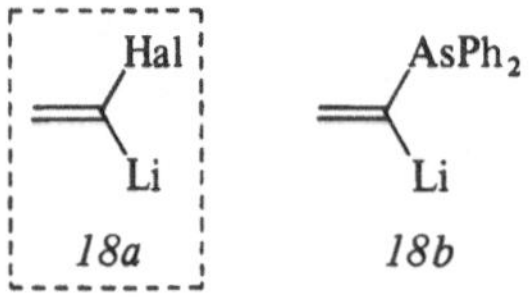

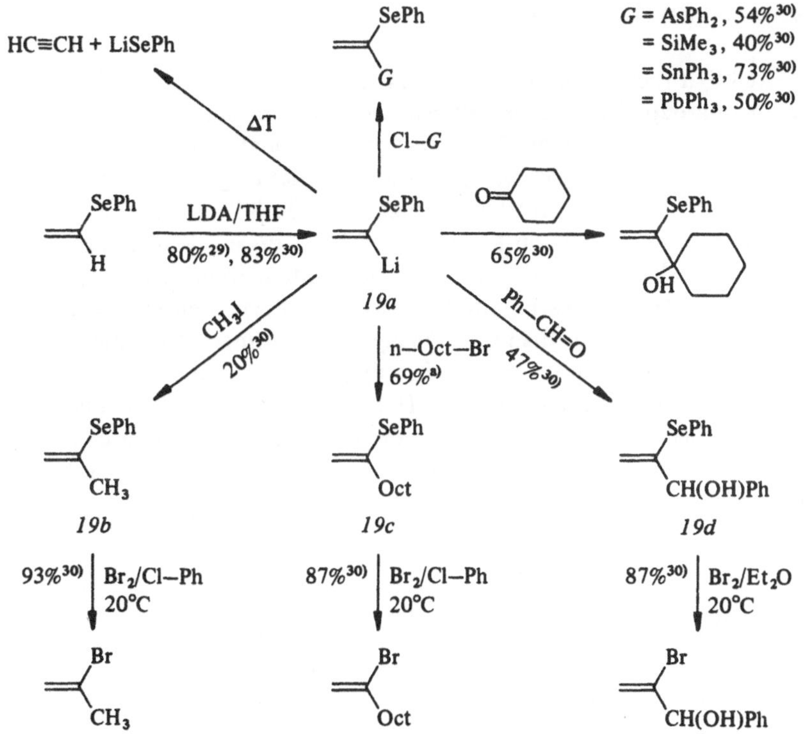

HC≡CH + LiSePh

G = AsPh$_2$, 54%[30]
  = SiMe$_3$, 40%[30]
  = SnPh$_3$, 73%[30]
  = PbPh$_3$, 50%[30]

ΔT

Cl–G

LDA/THF
80%[29], 83%[30]

65%[30]

SePh
Li

19a

CH$_3$I
20%[30]

n–Oct–Br
69%[a]

Ph–CH=O
47%[30]

OH

SePh
CH$_3$

19b

SePh
Oct

19c

SePh
CH(OH)Ph

19d

93%[30] | Br$_2$/Cl–Ph
20°C

87%[30] | Br$_2$/Cl–Ph
20°C

87%[30] | Br$_2$/Et$_2$O
20°C

Br
CH$_3$

Br
Oct

Br
CH(OH)Ph

Scheme 23 (a) This yield under conditions elucidated by Krief et al.[29] for the reaction of *19a* with decylbromid: HMPA/THF [1:20]. In THF the yield is only 15%.

and by us[30] independently — to lithiate phenylvinylselenium to give *19a* (Scheme 23).

Reacting the lithiation product *19a* with electrophiles gave the products stated in Scheme 23 in yields which are satisfactory considering the fact that this compound tends to split off lithiumphenylselenide[29, 30]. This α-elimination is favorized by addition of HMPA to the solvent. But a little HMPA is according to Krief[29] favorable for the reaction with alkyl halides. We have found that the selenium/bromine exchange on compounds *19b–19d* works very well by treating with bromine in chlorbenzene or ether at room temperature. The thermal decay of *19a* reminds to the formulated thermal decay of compounds *19e*[19] at which normally the residue in *trans* position to the halogen migrates (Fritsch-Buttenberg-Wiechell rearrangement).

$$\begin{array}{c} R \\ \diagdown \\ C=C \\ \diagup \quad \diagdown \\ R' \quad Hal \end{array} \quad \begin{array}{c} Li \\ \diagup \end{array} \xrightarrow[\text{–LiHal}]{} R-C\equiv C-R'$$

19e (Hal = Cl, Br)

3.4 Reagents for Carbonyl Haloolefination

It is to be expected that it can be useful to combine a halogen equivalent with an other organoelement group, for instance with a leaving group of proven value in carbonyl olefination reactions. In Scheme 24 we have the combination of an halogen equivalent and of either the leaving group typical for "Horner reagents" or the leaving group typical for "Peterson reagents" (see Chap. 5.1.). Both "combination reagents" — this term is often used in our laboratory, if 2 or 3 organoelement groups are connected with a small hydrocarbon unit — are able to transform carbonyl groups to halogenated olefines. The diphenylarsanyl-diphenylphosphoryl-methylpotassium (*20a*) gives essentially better yields than the diphenylarsanyl-trimethylsilyl-methyl-lithium (*20c*). If reacting 20a with benzaldehyde only the *trans*-olefine is produced, whereas with 20c the *cis*-isomer is a byproduct. The right hand side of the Scheme 24 informs about stereochemistry of the bromolysis of alkenylarsanes: No matter if the *trans*- or *cis*-product is the starting material, the *trans*-configurated bromide has been formed exclusively[31].

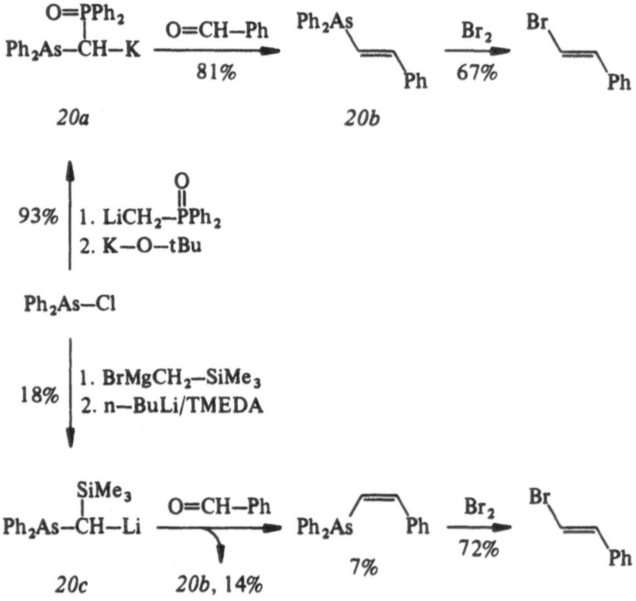

Scheme 24[31]

3.5 Strong Nucleophilic Equivalents for Br⁻ [32]

Scheme 25 shows that diphenyllithium- and diphenylpotassiumarsenide can serve as strong nucleophilic synthetic equivalents for the weak nucleophilic bromide ion (analogous possible for Cl⁻ and I⁻) which is not able to add to C-C-triple and C-C-double bonds.

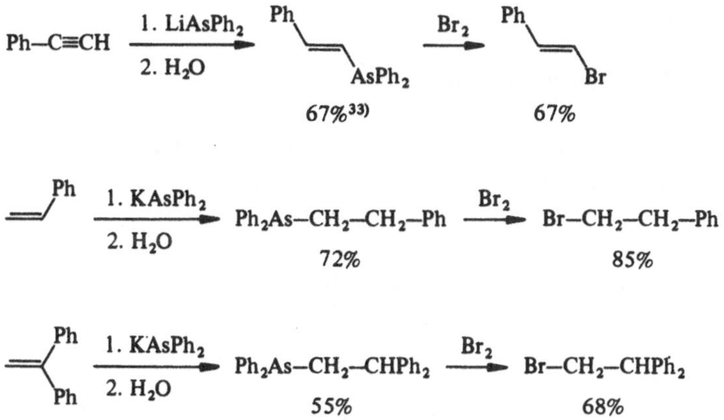

$$Ph-C\equiv CH \xrightarrow[\text{2. H}_2\text{O}]{\text{1. LiAsPh}_2} \underset{\substack{\text{AsPh}_2\\ 67\%^{33)}}}{\overset{Ph}{\diagup\!\!\!=\!\!\!\diagdown}} \xrightarrow{\text{Br}_2} \underset{\substack{\text{Br}\\ 67\%}}{\overset{Ph}{\diagup\!\!\!=\!\!\!\diagdown}}$$

$$\overset{Ph}{=\!\!\!\diagup} \xrightarrow[\text{2. H}_2\text{O}]{\text{1. KAsPh}_2} \underset{72\%}{Ph_2As-CH_2-CH_2-Ph} \xrightarrow{\text{Br}_2} \underset{85\%}{Br-CH_2-CH_2-Ph}$$

$$\overset{Ph}{\underset{Ph}{=\!\!\!<}} \xrightarrow[\text{2. H}_2\text{O}]{\text{1. KAsPh}_2} \underset{55\%}{Ph_2As-CH_2-CHPh_2} \xrightarrow{\text{Br}_2} \underset{68\%}{Br-CH_2-CHPh_2} \qquad \textbf{Scheme 25}$$

3.6 Reagent for Carbanion Halogenation

Whereas up to this point we have dealt with nucleophilic reagents, we shall now pass to electrophilic reagents which were applied by us to transform organolithium compounds according to Scheme 26.

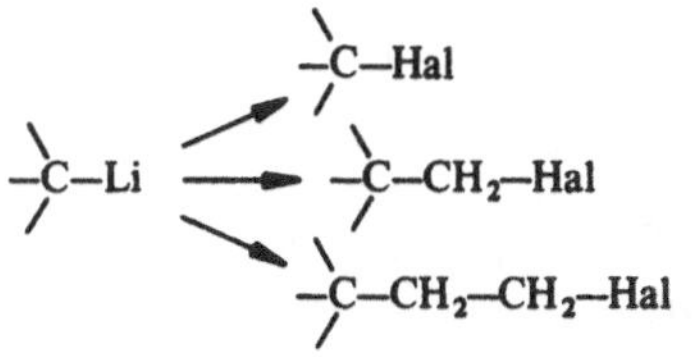

$$-\overset{|}{\underset{|}{C}}-Li \longrightarrow \begin{cases} -\overset{|}{\underset{|}{C}}-Hal \\ -\overset{|}{\underset{|}{C}}-CH_2-Hal \\ -\overset{|}{\underset{|}{C}}-CH_2-CH_2-Hal \end{cases} \qquad \textbf{Scheme 26}$$

If one tries to transfer carbanions to halogen compounds by reacting with elementary halogen oxidative coupling often is the main reaction. Better results are observed with carbontetrachloride, carbontetrabromide, or other polyhaloalkenes as reagents[34]. As a good alternative we offer the reaction with diphenylarsanechloride and then with bromine, iodine, or sulfurylchloride as has been demonstrated by the example shown in Scheme 27.

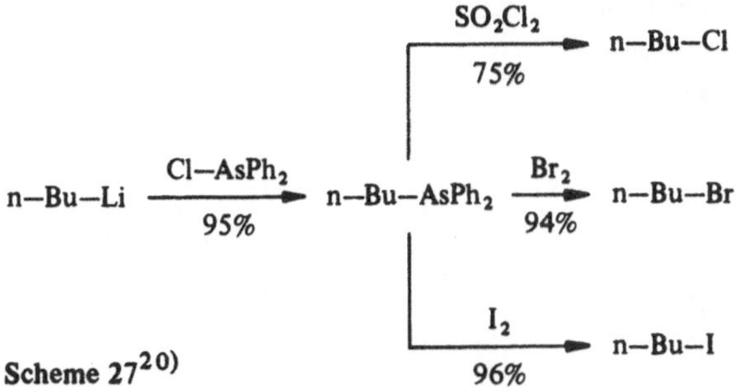

$$n-Bu-Li \xrightarrow[95\%]{Cl-AsPh_2} n-Bu-AsPh_2 \begin{cases} \xrightarrow[75\%]{SO_2Cl_2} n-Bu-Cl \\ \xrightarrow[94\%]{Br_2} n-Bu-Br \\ \xrightarrow[96\%]{I_2} n-Bu-I \end{cases}$$

Scheme 27[20)]

127

3.7 Reagents for Carbanion Chain Elongation

a) Diphenylvinylarsane and Phenylvinylselenide

Chain extension of organo lithium compounds according to Scheme 28 are precluded because the first step is impossible.

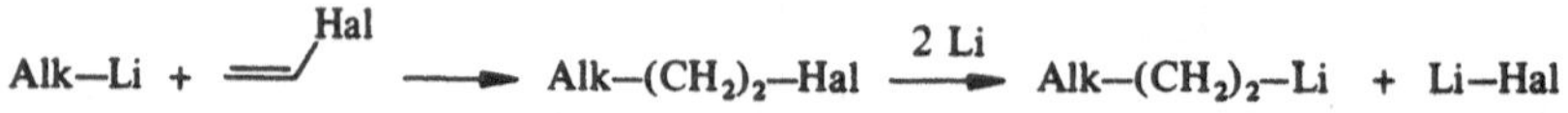

$$\text{Alk—Li} + \overset{\text{Hal}}{=\!\!/} \longrightarrow \text{Alk—(CH}_2)_2\text{—Hal} \xrightarrow{\text{2 Li}} \text{Alk—(CH}_2)_2\text{—Li} + \text{Li—Hal}$$

Scheme 28

The addition of lithioalkanes to diphenylvinylarsane according to Scheme 29 constitutes an equivalent reaction, for the diphenylarsanyl group of the resulting primary alkyl-diphenyl-arsane is readily replaceable by halogens which are in turn replaceable by lithium in a well-known reaction[10d, 35].

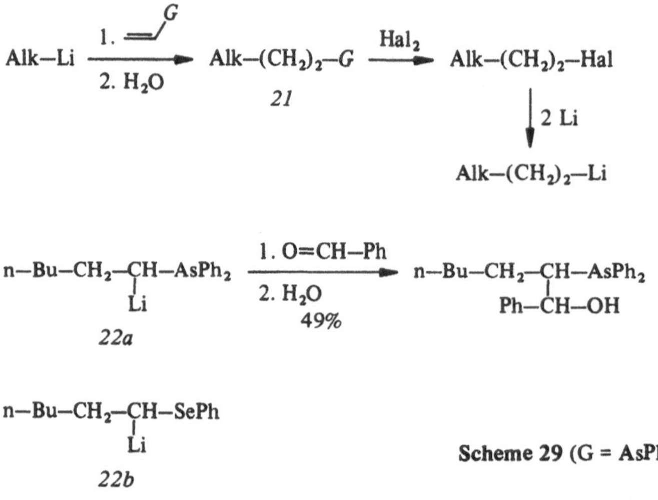

$$\text{Alk—Li} \xrightarrow[\text{2. H}_2\text{O}]{\text{1. } =\!\!/^G} \underset{21}{\text{Alk—(CH}_2)_2\text{—}G} \xrightarrow{\text{Hal}_2} \text{Alk—(CH}_2)_2\text{—Hal}$$

$$\downarrow \text{2 Li}$$

$$\text{Alk—(CH}_2)_2\text{—Li}$$

$$\underset{22a}{\text{n—Bu—CH}_2\text{—}\underset{\text{Li}}{\text{CH}}\text{—AsPh}_2} \xrightarrow[\substack{\text{2. H}_2\text{O} \\ 49\%}]{\text{1. O=CH—Ph}} \text{n—Bu—CH}_2\text{—}\underset{\text{Ph—CH—OH}}{\text{CH}}\text{—AsPh}_2$$

$$\underset{22b}{\text{n—Bu—CH}_2\text{—}\underset{\text{Li}}{\text{CH}}\text{—SePh}}$$

Scheme 29 (G = AsPh$_2$, SePh)

Table 6. Yield of product *21*

G	Alk	Yield (%)
	C$_2$H$_5$	37
	n-C$_4$H$_9$	95
—AsPh$_2$	s-C$_4$H$_9$	71
	t-C$_4$H$_9$	57
	n-C$_4$H$_9$	72
—SePh	s-C$_4$H$_9$	25
	t-C$_4$H$_9$	25

We also succeeded in adding alkyl lithium compounds to phenylvinylselenide[10a, 35]. However the arsenic reagent is greatly superior to phenylvinylselenide as an electrophilic chain elongation reagent since poorer yields are obtained, both in the addition step (Table 6) and in the halogenolysis step. — Krief et al.[29] have confirmed our experiment with n-butyllithium and succeeded in using the anionic adduct *22b* as a nucleophilic reagent. Analogous reactions with the corresponding adduct *22a* of diphenylvinylarsane were performed by us[10a] (Scheme 29).

b) Chlor-diphenylarsanyl-methane (*23a*, n = 1) and 1-Chlor-2-diphenylarsanyl-ethane (*23a*, n = 2) and Analogous Selenium Compounds

In theory a chain elongation of organolithium compounds with an arbitrary number of carbon atoms should be possible by substitution reactions with reagents of the type *23a* and *23b*. Such reagents are obtainable according to Scheme 30[36]. With

$$Cl-(CH_2)_n-Cl \ + \ Li-AsPh_2 \ \longrightarrow \ Cl-(CH_2)_n-AsPh_2 \qquad 23a$$
$$[\text{Yield: n = 1, 2, 3, 4: 32, 42, 43, 41\%}]$$

$$Cl-(CH_2)_n-Cl \ + \ Na-SePh \ \longrightarrow \ Cl-(CH_2)_n-SePh \qquad 23b$$
$$[\text{Yield: n = 1, 2, 3, 4, 6: 45, 69, 77, 87, 92\%}] \qquad \textbf{Scheme 30}$$

n = 1 or 2 they reacted well with n-butyllithium. Bromolysis of the products afforded the corresponding bromides in satisfactory yields. The yield over all in both cases was better with the arsenic compounds[36] (Scheme 31). With reagents *23a* and *23b*, n > 2 surprisingly no substitution occurred and we could recover the chlorides nearly quantitatively from the reaction mixture[36]. The obvious explanation that the methylene group in α-position to the chlorine is lithiated giving a chelate stabilized

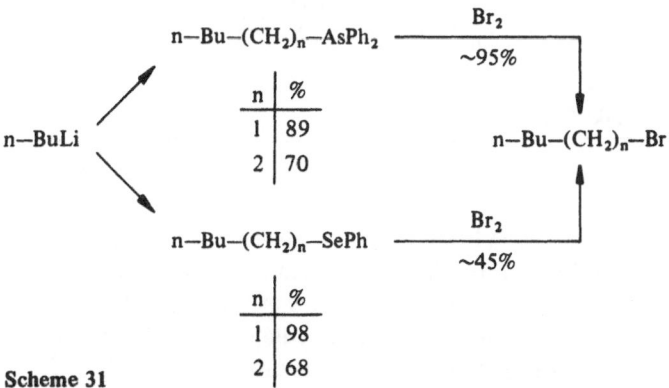

Scheme 31

lithium compound of the type *23c*, has been ruled out by deuteration experiments. So the surprising stability of compounds *23a* and *23b*, n > 2 towards n-butyllithium presently is still mysterious.

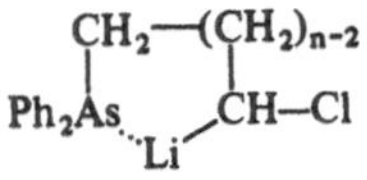

$$\begin{array}{c}
CH_2{-}(CH_2)_{n-2}\\
Ph_2As\quad\quad CH{-}Cl\\
{}^{\,\cdots}Li
\end{array}$$

23c

4 Element/Lithium Exchange in Organic Synthesis

4.1 General Remarks

This chapter deals with the replacement of organoelement groups by a lithium atom. We have taken an interest in this mode of reaction when trying to prepare the forementioned reagent diphenylarsanyl-methyllithium (*9*). During these experiments the author got the impression that element/lithium exchange offers many unexploited possibilities in organic synthesis. A general remark about this reaction type seems advisable at this point.

Element/lithium exchange has found application in organic synthesis nearly exclusively with tin and selenium as the heteroatoms. The work was mainly done by the groups of Seyferth, Seebach, and Krief. In Scheme 32 some examples are

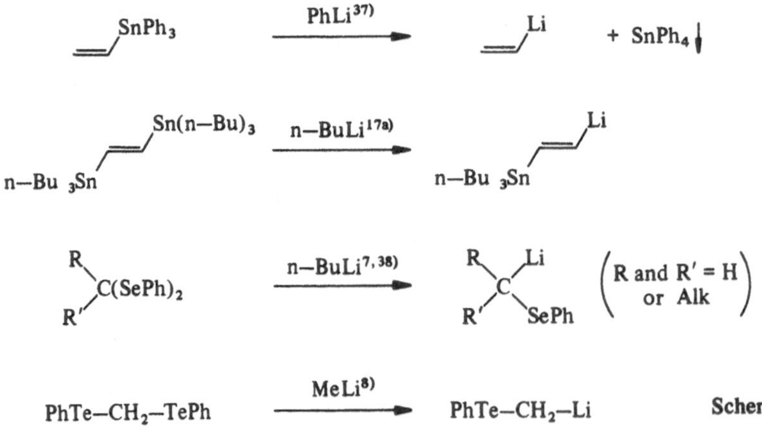

$$\underset{}{\diagup}SnPh_3 \quad \xrightarrow{\;PhLi^{37)}\;} \quad \underset{}{\diagup}Li \quad + SnPh_4\downarrow$$

$$n{-}Bu_3Sn\diagup\overset{Sn(n{-}Bu)_3}{\diagup} \quad \xrightarrow{\;n{-}BuLi^{17a)}\;} \quad n{-}Bu_3Sn\diagup\overset{Li}{\diagup}$$

$$\underset{R'}{\overset{R}{>}}C(SePh)_2 \quad \xrightarrow{\;n{-}BuLi^{7,\,38)}\;} \quad \underset{R'}{\overset{R}{>}}\underset{SePh}{\overset{Li}{C}}\quad\left(\begin{array}{c}\text{R and R}' = \text{H}\\ \text{or Alk}\end{array}\right)$$

$$PhTe{-}CH_2{-}TePh \quad \xrightarrow{\;MeLi^{8)}\;} \quad PhTe{-}CH_2{-}Li \qquad \text{Scheme 32}^{7,\,8,\,37,\,38)}$$

given. But many observations, scattered in literature, show clearly that reactions of this type are possible in the whole area of the elements stated in Scheme 4. According to Wittig and Maercker[39] who reacted phenylderivatives of various elements with p-tolyllithium (Scheme 33), ate-complexes are intermediates in those exchange reactions, and the exchange rate was found increasing in the order As < P ≪ Sb < Bi.

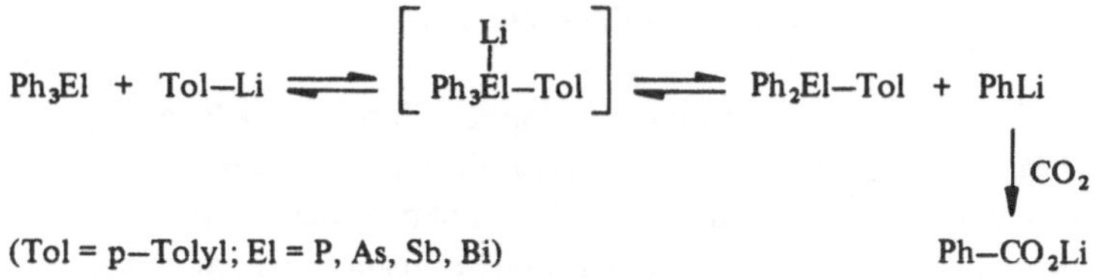

$$\text{Ph}_3\text{El} + \text{Tol}\text{--Li} \rightleftharpoons \left[\begin{array}{c} \text{Li} \\ | \\ \text{Ph}_3\text{El}\text{--Tol} \end{array} \right] \rightleftharpoons \text{Ph}_2\text{El}\text{--Tol} + \text{PhLi}$$

$$\downarrow \text{CO}_2$$

(Tol = p—Tolyl; El = P, As, Sb, Bi) $\text{Ph--CO}_2\text{Li}$ Scheme 33[39]

The expression "element/lithium exchange" has been proposed[40] by us in analogy to the expression "halogen/lithium exchange" and in order to separate this important reaction type from the numberless transmetalation reactions.

4.2 Synthesis of Reagents for Nucleophilic Halomethylation

a) Diphenylarsanyl-methyllithium 9

Lithiomethyl-diphenyl-arsaneoxide (10) is a good synthetic equivalent for the synthons Li-CH$_2$-Hal (Chap. 3), but a simpler equivalent would be the hitherto unknown compound 9. We succeeded in preparing 9 by treating 24 with n-butyllithium in high excess (4 moles; Scheme 34). The remaining n-butyllithium was destroyed by reacting 1 hour with tetrahydrofurane at room temperature (formation of ethylene and the lithiumenolate of acetaldehyde). The new reagent attacks electrophilic compounds[40, 44] yielding products (Table 7) which we had received earlier by reduction of corresponding arsaneoxides. With this reagent one saves the reduction step, but the corresponding oxide is better accessible. Note that arsenic/lithium exchange occurs not quantitative because phenyl/butyl exchange is a side reaction.

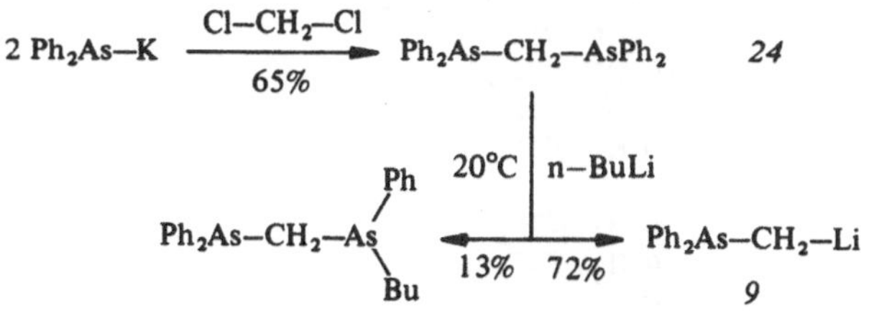

$$2\ \text{Ph}_2\text{As}\text{--K} \xrightarrow[65\%]{\text{Cl--CH}_2\text{--Cl}} \text{Ph}_2\text{As}\text{--CH}_2\text{--AsPh}_2 \qquad 24$$

$$\text{Ph}_2\text{As}\text{--CH}_2\text{--As} \overset{\text{Ph}}{\underset{\text{Bu}}{\diagdown}} \xleftarrow[13\%]{} \overset{20°C\ |\ n\text{--BuLi}}{\underset{72\%}{\xrightarrow{}}} \text{Ph}_2\text{As}\text{--CH}_2\text{--Li}$$

                                                                      9   Scheme 34[40]

Table 7. Reactions of diphenylarsanyl-methyllithium 9 with electrophilic compounds[44]

Reactant	Product Ph$_2$As--CH$_2$--E	Yield (%)
1-Iodopropane	E = --(CH$_2$)$_2$CH$_3$	72
1-Bromobutane	--(CH$_2$)$_3$CH$_3$	65
1-Bromooctane	--(CH$_2$)$_7$CH$_3$	44
Benzaldehyde	--CH(OH)Ph	69
Butyraldehyde	--CH(OH)(CH$_2$)$_2$CH$_3$	62
Cyclohexanone	--C(OH) (CH$_2$)$_5$	70

b) Diphenylstibanyl-methyllithium *25a* and -methylcopper (I) *25b*

The unwelcome phenyl/butyl exchange observed during reaction *24* ⟶ *9* can not occur during the synthesis of diphenylstibanyl-methyllithium *25* according to Scheme 35. Owing to the higher electrophilicity of the diphenylstibanyl group, element/lithium exchange in this case is possible with phenyllithium which is less reactive than n-butyllithium. But the new reagent which has been formed quantitatively proved to be distinctly less reactive than the arsenic analogon since reacting only very slowly with alkylhalides. Nevertheless we obtained a reagent, reactive to alkylhalides, by transmetalation to the corresponding copper(I) compound *25b*. Yields of alkylation products obtained by reacting with *25b* are not extremely high but acceptable[40, 44].

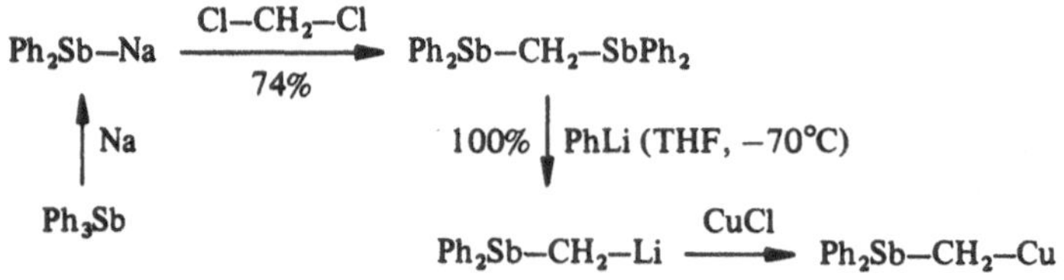

$$Ph_2Sb-Na \xrightarrow[74\%]{Cl-CH_2-Cl} Ph_2Sb-CH_2-SbPh_2$$

$$\uparrow Na \qquad\qquad 100\% \downarrow PhLi\ (THF,\ -70°C)$$

$$Ph_3Sb \qquad\qquad Ph_2Sb-CH_2-Li \xrightarrow{CuCl} Ph_2Sb-CH_2-Cu$$

Scheme 35[40)]   *25a*   *25b*

Table 8. Reactions of diphenylstibanyl-methyllithium *25a* and -methylcopper *25b* with electrophilic compounds[44)]

Reagent	Reactant	Product Ph_2Sb-CH_2-E	Yield (%)
	Benzaldehyde	E = $-CH(OH)Ph$	39
	Butyraldehyde	$-CH(OH)(CH_2)_2CH_3$	66
	Benzophenone	$-C(OH)Ph_2$	45
25a	Cyclohexanone	$-\overline{C(OH)(CH_2)_5}$	61
	1-Bromopropane	$-(CH_2)_2CH_3$	12
	1-Iodopropane	$-(CH_2)_2CH_3$	14
	1-Iodopropane	$-(CH_2)_2CH_3$	80
25b	1-Iodohexane	$-(CH_2)_5CH_3$	45
	1-Iodooctan	$-(CH_2)_7CH_3$	61

4.3 Synthesis of Reagents for Nucleophilic Lithiomethylation and Carbonyl Olefination

Starting from compounds $Ph_3M-CH_2-MPh_3$ we were able to prepare by element/lithium exchange the reagents triphenylstannyl-methyllithium (27[15d, 23, 41)]) and triphenylplumbyl-methyllithium (28[15d, 23, 40)]) (Scheme 36). In case of the tin derivative element/lithium exchange is not a realistic alternative to the access by halogen/lithium exchange. But in the case of the lead derivative this method is superior to

$(Ph_3Ge)_2CH_2$ $\xrightarrow{\quad RLi^{42a)}\quad}\!\!/\!\!/\!\!\longrightarrow$ Ph_3Ge-CH_2-Li $\xleftarrow[85\%]{\quad n-BuLi^{43)}\quad}$ Ph_3Ge-CH_2-Br

26

$(Ph_3Sn)_2CH_2$ $\xrightarrow[36\%]{\quad PhLi\quad}$ Ph_3Sn-CH_2-Li $\xleftarrow[98\%]{\quad n-BuLi\quad}$ Ph_3Sn-CH_2-I

27

$(Ph_3Pb)_2CH_2$ $\xrightarrow[\sim100\%]{\quad PhLi\quad}$ Ph_3Pb-CH_2-Li $\xleftarrow[58\%]{\quad n-BuLi\quad}$ Ph_3Pb-CH_2-I

28

Scheme 36

the other one. The germanium compound *26*, in contrast to *27* and *28* described in literature[43], was obtained by halogen/lithium exchange only. In the following section we shall get to know a second method for the preparation of this compounds. Whereas the germanium compound *26* is very reactive[41b], the reagents *27* and *28* show diminished reactivity. They are able to add to carbonyl compounds at which the yields are rather modest (Table 9). But both compounds failed to react with alkylhalides, and in contrary to our findings with the analogous antimony species transmetalation to the copper compound did not change the situation[44]. Altogether we can distinguish two groups of reagents with respect to the reactivity against alkylhalides. The reagents with the lighter organoelement groups are reactive, the others are not (Scheme 37).

Reaction with alkylhalides:

Ph_3Ge-CH_2-Li $\qquad\qquad$ Ph_2As-CH_2-Li $\qquad\qquad$ $PhSe-CH_2-Li$

No reaction with alkylhalides:

Ph_3Sn-CH_2-Li $\qquad\qquad$ Ph_2Sb-CH_2-Li $\qquad\qquad$ $PhTe-CH_2-Li$
Ph_3Pb-CH_2-Li

Scheme 37

Table 9. Reactions of triphenylstannyl-methyllithium *27* and triphenylplumbyl-methyllithium *28* with carbonyl compounds[44]

Reagent	Electrophile	Product Ph_3El-CH_2-E	Yield (%)
27	Benzaldehyde	$E = -CH(OH)Ph$	65
	Butyraldehyde	$-CH(OH)(CH_2)_2CH_3$	31
	Acetophenone	$-CPh(OH)CH_3$	41
	Cyclohexanone	$-\overline{C(OH)(CH_2)_5}$	37
28	Benzaldehyde	$-CH(OH)Ph$	39
	Benzophenone	$-C(OH)Ph_2$	45
	Cyclohexanone	$-\overline{C(OH)(CH_2)_5}$	46

133

Our results seem to be instructive in account of the following reasons: the halogen/lithium exchange and the element/lithium exchange are equilibrium processes leading to the compound of highest thermodynamic stability, if the equilibrium — as in our experiments — can adjust undisturbed.

Hence it has been demonstrated that the heavy heteroatoms are stabilizing a negative charge in α-position. Moreover the low reactivity of the lithium compounds with heavy organoelement groups points to an especially good stabilization of the negative charge. So one can assume that we have similar conditions as with halogenated carbanions[45] (Scheme 38) where the increasing acidity by passing from fluorine to iodine compounds points to increasing stabilization of the negative charge in the same direction. Whereas the increase in acidity by passing from fluorine to chlorine can be rationalized in terms of (p⟶d) π-stabilization of the anion, the particular strong stabilizing effect of bromine and iodine needs another explanation because of the longer carbon halogen bonds (see l.c.[6]).

Later on in this treatise (Chap. 6) hydrogen/lithium exchange at substrates with heavy organoelement groups is reported on. These reactions are showing once more that such groups are significantly electron withdrawing.

F_3CH $\quad$ Cl_3CH $\quad\quad$ Br_3CH $\longrightarrow$ $\quad$ I_3CH

$\quad\quad\quad$ Increasing acidity

Stabilization of negative charge in $Hal-C\ominus$:

$\quad\quad\quad$ $F < Cl < Br \sim I$ $\quad\quad\quad\quad\quad\quad\quad$ **Scheme 38**

What about the applicability of organolithium compounds of type *29* with heavy organoelement groups in organic synthesis? At present we can realize three modes of use (Scheme 39). The antimony compound is suitable as a reagent for nucleophilic halomethylation[21]. Nucleophilic lithiomethylation and carbonyl olefination, two other modes of application, are the objects of Sects. 4.4, 4.5, and Chap. 5.

$\quad\quad\quad\quad$ Equivalent for $Hal-CH_2-Li$

$G-CH_2-Li \equiv$ Equivalent for $Li-CH_2-Li$

29 $\quad\quad$ Carbonyl olefination $\quad\quad\quad\quad\quad$ **Scheme 39**

4.4 Nucleophilic Lithiomethylation

All mentioned compounds of the type *29* with heavy organoelement groups (element = Sn, Pb, Sb) are equivalents for dilithiummethane and reagents for nucleophilic lithiomethylation. In Scheme 40 two examples are given: By reacting with the lead

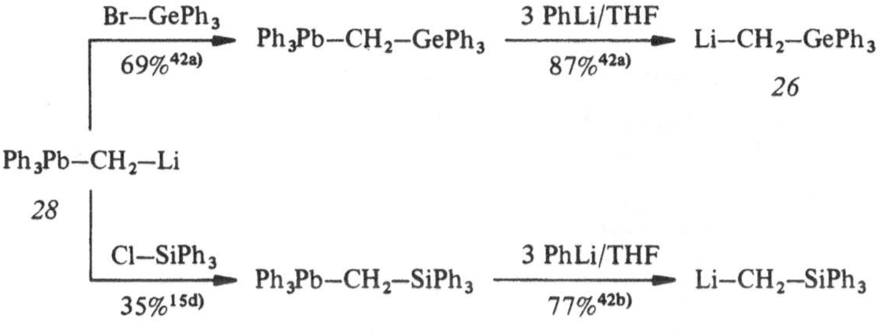

$$\text{Ph}_3\text{Pb}-\text{CH}_2-\text{Li} \quad \xrightarrow[\underset{69\%^{42a)}}{\text{Br}-\text{GePh}_3}]{} \quad \text{Ph}_3\text{Pb}-\text{CH}_2-\text{GePh}_3 \quad \xrightarrow[\underset{87\%^{42a)}}{\text{3 PhLi/THF}}]{} \quad \text{Li}-\text{CH}_2-\text{GePh}_3$$

$$26$$

$$\text{Ph}_3\text{Pb}-\text{CH}_2-\text{Li}$$

$$28$$

$$\xrightarrow[\underset{35\%^{15d)}}{\text{Cl}-\text{SiPh}_3}]{} \quad \text{Ph}_3\text{Pb}-\text{CH}_2-\text{SiPh}_3 \quad \xrightarrow[\underset{77\%^{42b)}}{\text{3 PhLi/THF}}]{} \quad \text{Li}-\text{CH}_2-\text{SiPh}_3$$

Scheme 40

compound *28* triphenylgermylbromide and triphenylsilylchloride are transformed to the corresponding methyllithium derivative (for nucleophilic lithioalkylation with selenium reagents see l. c.[38c)]).

4.5 Nucleophilic Bislithiomethylation

By reason of these applicabilities corresponding reagents with 2 or 3 heavy organo-element groups seem to be attractive as well. Having at first vainly tried to prepare compounds of this type by hydrogen/lithium exchange on the corresponding methylene compounds G-CH$_2$-G, we synthesized the tri- and tetrasubstituted methanes of Scheme 41 (all described in literature) and reacted them with phenyllithium or n-butyllithium.

(Ph$_2$Sb)$_3$CH	(Me$_3$Sn)$_4$C
30a	
(Ph$_3$Pb)$_3$CH	(Ph$_3$Pb)$_4$C
30b	**Scheme 41**

The *tetra*substituted methanes turned out to be completely inert presumably owing to steric hindrance of base addition to the heteroatoms. The trisubstituted antimony compound *30a* was inert too. In this case the reason may be that *30a* is a very strong threedentate ligand which desactivates organolithium compounds by complexation. But the desired exchange of an organoelement group for lithium occurred smoothly[46)] at the well accessible[47)] tris(triphenylplumbyl)methane *30b* which therefore seems to be a useful substance.

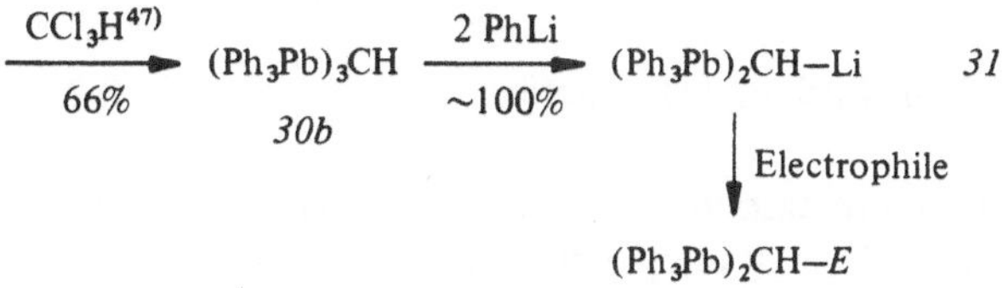

$$\text{Ph}_3\text{Pb}-\text{Li} \quad \xrightarrow[\underset{66\%}{\text{CCl}_3\text{H}^{47)}}]{} \quad \underset{30b}{(\text{Ph}_3\text{Pb})_3\text{CH}} \quad \xrightarrow[\underset{\sim100\%}{\text{2 PhLi}}]{} \quad (\text{Ph}_3\text{Pb})_2\text{CH}-\text{Li} \quad 31$$

$$\downarrow \text{Electrophile}$$

Scheme 42

$$(\text{Ph}_3\text{Pb})_2\text{CH}-E$$

Table 10. Reactions of bis(triphenylplumbyl)methyllithium *31* with electrophiles[46)]

Electrophile	Product $(Ph_3Pb)_2CH-E$	Yield (%)
Benzaldehyde	$E = -CH(OH)Ph$	86
Butyraldehyde	$-CH(OH)C_3H_7$	62
$Cl-SiMe_3$	$-SiMe_3$	70
$Br-GePh_3$	$-GePh_3$	87
$Cl-SnPh_3$	$-SnPh_3$	73
$Cl-AsPh_2$	$-AsPh_2$	65

30b reacts with an excess of phenyllithium to the wished-for compound *31* which proved to be surprisingly reactive against carbonyl compounds (Table 10) and organo-element halides, whereas with alkylhalides no reaction occurred[46)].

Some of the products are starting materials for interesting combination reagents. In compound *32a* and *32b* we have a lithium equivalent together with a good leaving group for a carbonyl olefination reaction and in *32c* is a lithium equivalent combined with an halogen equivalent. Compound *32d* with a tin atom and two lead atoms offers an interesting competition phenomenon. As one can see from the yields (Scheme 43[46)]) the two different organoelement groups suffer exchange for lithium with approximately the same rate.

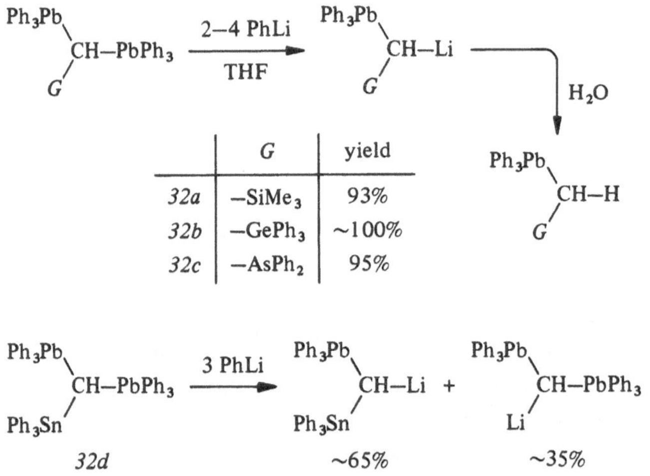

	G	yield
32a	$-SiMe_3$	93%
32b	$-GePh_3$	~100%
32c	$-AsPh_2$	95%

Scheme 43[46)]

5 New Carbonyl Olefination Reagents

5.1 General Remarks about Carbonyl Olefination

The detection of olefination of carbonyl compounds with phosphoranes by Wittig[48a)] in 1952 stimulated the development of organic synthesis in a high degree, though in contrast to his habitual behavior Wittig did not investigate this important reaction systematically.

After the discovery of the Wittig reaction and the modification according to Horner[48b] sulfur compounds[49] and arsenic compounds[50] were checked up with regard to corresponding ability, but epoxides, epoxide/olefine mixtures or other products were obtained instead of pure olefines. More favorable turned out to be the neighbor element of phosphorus at the left hand side in the periodic table, the silicium. As explored by Peterson[51a] 16 years after Wittigs detection, carbonyl compounds are smoothly olefinated according to Scheme 44. By reason of the high

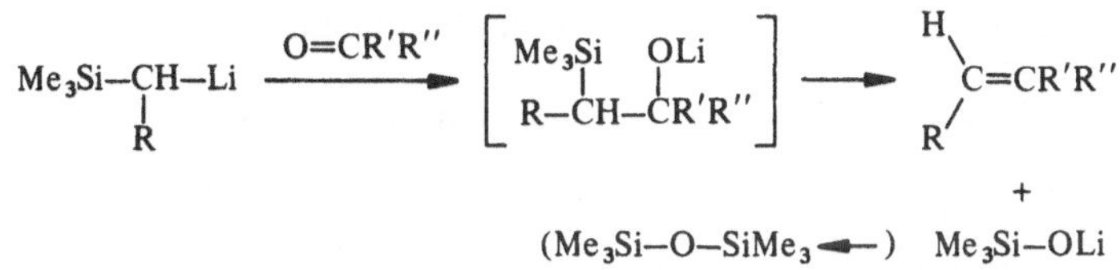

$$Me_3Si-\underset{\underset{R}{|}}{CH}-Li \xrightarrow{O=CR'R''} \left[\begin{matrix} Me_3Si & OLi \\ | & | \\ R-CH-CR'R'' \end{matrix} \right] \longrightarrow \underset{R}{\overset{H}{\diagdown}}C=CR'R''$$

$$+$$

$$(Me_3Si-O-SiMe_3 \longleftarrow) \quad Me_3Si-OLi$$

Scheme 44   *33*

nucleophilicity of the silicium reagent and of the volatility of the byproduct *33* this type of carbonyl olefination occasionally is preferred to the Wittig type. Three years ago Schrock[51b] discovered the possibility of carbonyl olefination with tantalum- and niobium-ylides; such reagents are able to olefinate even carboxylic esters, amide, and CO_2.

There is no real need today for new carbonyl olefination methods as such. But though some techniques have been developed for highly stereoselective carbonyl olefination[52] further progress in this direction would be welcome.

5.2 Carbonyl Olefination Reagents with Heavy Organoelement Groups as Leaving Groups

Among the various alcohols, obtained by us by reacting new reagents with carbonyl compounds, there are two types which are said to undergo a smooth elimination reaction in acidic solution: Stimulated by some previous findings Davis and Gray[53] have demonstrated that alcohols with a triphenylstannyl- or triphenylplumbyl group in β-position to the hydroxy group, prepared by nucleophilic opening of epoxides (Scheme 45), suffer quick decomposition in acidic solution at room temperature. The reaction over all proceeds with complete retention. Because opening of the oxirane ring certainly occurs by backside attack, the β-elimination is an anti-elimina-

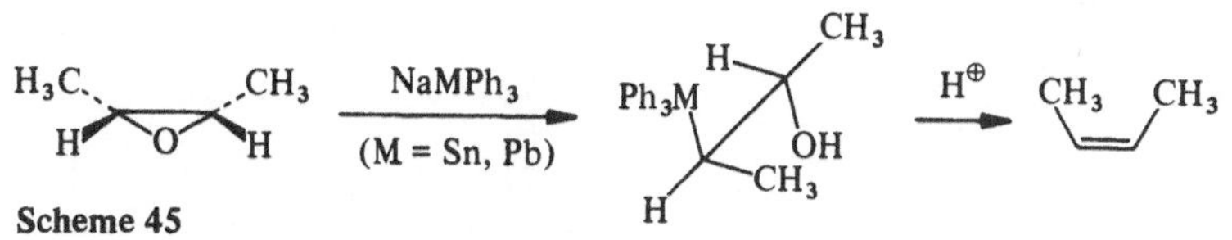

$$\underset{H}{\overset{H_3C}{\diagdown}}\underset{O}{\overset{}{\triangle}}\underset{H}{\overset{CH_3}{\diagup}} \xrightarrow[\text{(M = Sn, Pb)}]{NaMPh_3} Ph_3M \underset{H}{\overset{\overset{\displaystyle CH_3}{\underset{H}{\diagup}}}{\diagup}}\underset{CH_3}{\overset{OH}{\diagdown}} \xrightarrow{H^{\oplus}} \underset{}{\overset{CH_3 \quad CH_3}{\diagup\diagdown}}$$

Scheme 45

137

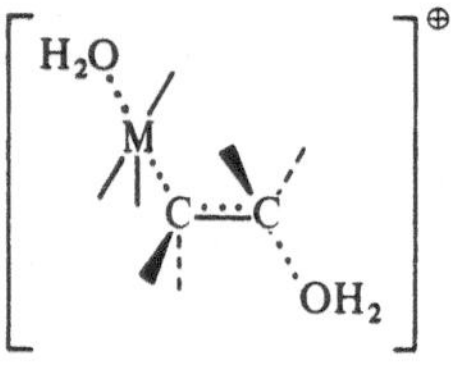

34

tion. Hence the transition state *34* was postulated for the rate determining step of the second order reaction[53]. Combination of this elimination induced by acid with our carbon-carbon connective synthesis of alcohols of the type *35a* (Chap. 4) has led in our laboratory[54] to new methods of carbonyl olefination (Scheme 46, method A; Table 11).

Four lithium derivatives which are accessible by element/lithium exchange (Chap. 4) triphenylgermyl- *26*, triphenylstannyl- *27*, triphenylplumbyl- *28* and diphenylstibanyl-methyllithium *25a*, proved to be carbonyl olefination reagents. In the right hand side column of Table 11 the yield is given, related to compound *29*. The alcohols *35a* have not been isolated in these experiments. The simplest procedure

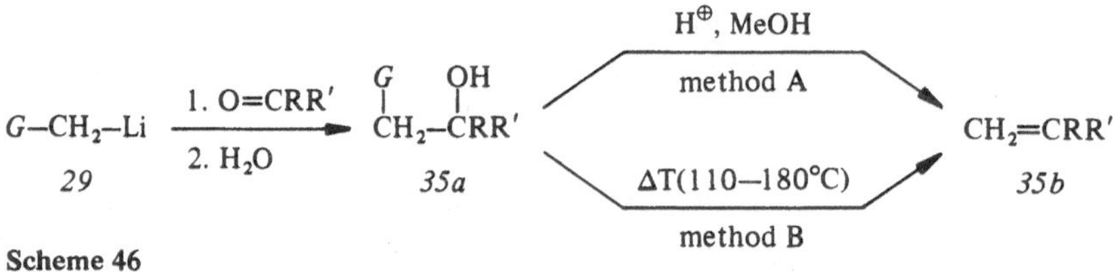

Scheme 46

Table 11. Carbonyl olefination according Scheme 46, method A[54] (elimination of G–OH by excess of perchloric acid in methanol at ~ 20 °C)

G	R	R′	Yield of Olefine (%)	
			$35a \rightarrow 35b$	$29 \rightarrow 35b$
Ph$_3$Ge–	H	Ph	95	
	CH$_3$	Ph	99	
	H	Ph	68	80
Ph$_3$Sn–	H	–(CH$_2$)$_2$CH$_3$	63	
	–(CH$_2$)$_5$–		89	78
	H	Ph	77	67
Ph$_3$Pb–	Ph	Ph	98	
	–(CH$_2$)$_5$–		80	75
	H	Ph	71	61
Ph$_2$Sb–	Ph	Ph	61	
	–(CH$_2$)$_5$–		21	42

Table 12. Carbonyl olefination according Scheme 46, method B[54]

G	R	R'	Yield of olefine (%)		Temp.
			$35a \rightarrow 35b$	$29 \rightarrow 35b$	°C
Ph_3Ge-	H	Ph	0		180
	CH_3	Ph	34		
	H	Ph	84	87	110
	H	$-(CH_2)_2CH_3$	100	75	130
Ph_3Sn-	H	$-(CH_2)_5CH_3$	70		130
	CH_3	Ph	71		130
	$-(CH_2)_5-$		96	86	130
	H	Ph	51		110
Ph_3Pb-	Ph	Ph	93	85	180
	$-(CH_2)_5-$		65		180
	H	Ph	48		180
Ph_2Sb-	Ph	Ph	68		180
	$-(CH_2)_5-$		37		110

for accomplishing the elimination reaction is to give the THF solution of the alcohol through a short column filled with silicagel[23, 55]. In addition to this we have found that decomposition of the alcohols can be achieved in neutral medium by heating (Scheme 46, method B; Table 12). This second method did not work with the germanium derivatives. The lead and antimony derivatives need higher temperature for decomposition (Table 12) and the yields are on an average slightly worse compared with the tin reagent.

Alcohols with organoelement groups listed in Table 13 gave with one exception only small amounts of olefine. But with β-hydroxyalkyl-selenides (*35a*, G = −SePh or −SeCH$_3$) stereospecific *trans*-elimination can be achieved in acidic (for instance excess of perchloric acid ether at room temperature) or basic media to give olefines in good yield[56]. So we can state that preparatively useful carbonyl olefination reactions in which epoxides are not a by-product, are allowed not only with phosphorus and silicium containing reagents but are possible in the wide area of the periodical table marked in Scheme 55c with little lines.

Table 13. Low-yield-carbonyl-olefinations according Scheme 46, method A and B[53]

Alcohols *35a*	Yield of styrene (%)	
G =	$H^\oplus$/MeOH	ΔT
Ph_2P-	8	8
Ph_2As-	15	45
PhS−	0	0
PhTe−	10	15

5.3 Stereospecific Carbonyl Olefination
with Phenylthiotriphenylstannyl-methyllithium *36*

We have studied these two methods with respect to the stereochemistry taking the sulfur-tin-derivative *36* as a reagent. Two diastereomeric alcohols (Scheme 47) were formed in the ratio of about 1:3 at which the higher melting isomer *37a* is the main product. Whereas *heating* of this compound gives practically pure *cis*-olefine, *decomposition in acidic solution* led to the *trans*-olefine only. So the addition is stereoselective, whereas the two modes of elimination are stereospecific. Because the lower melting isomer *37b*, as in the known reaction of Scheme 45 (M = Sn), gives by treating with acid the *cis*-olefine an analogous configuration (threo) can be assumed. The possibility of transforming each of the isomeric alcohols either to the *trans*- or *cis*-olefine allows the *stereospecific synthesis* of these two olefines starting from benzaldehyde. Further stereospecific olefine syntheses with reagents analogous to 36 are to the expected.

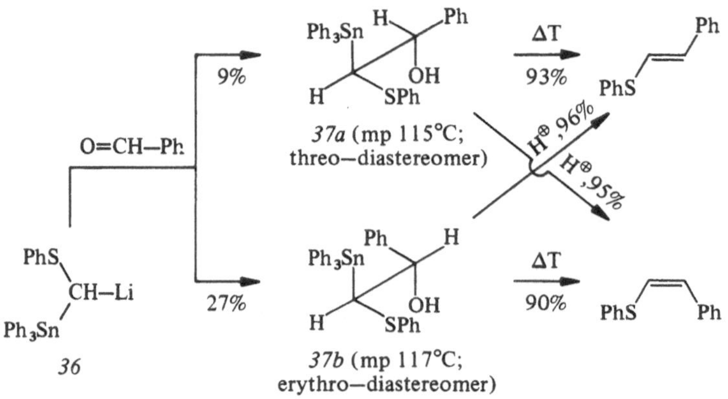

Scheme 47 (For reason of simplification only one of the two enantiomers is formulated in each case)

5.4 Plumbylolefination, Lithioolefination, and Phosphorylolefination of Aldehydes

With respect to special carbonyl olefinations now reagents merit an interest which contain more than one organoelement group. For instance it seems worth knowing if with the reagent bis(triphenylplumbyl)methyllithium (*31*) lithioolefination according to Scheme 48 is possible.

As reported in Chap. 4 reagent *31* is well accessible. It reacts with carbonyl compounds yielding β-bis(triphenylplumbyl)alcohols. But these compounds revealed to be surprisingly stable against acids or heat[65]. Apparently the carbon-lead bond is

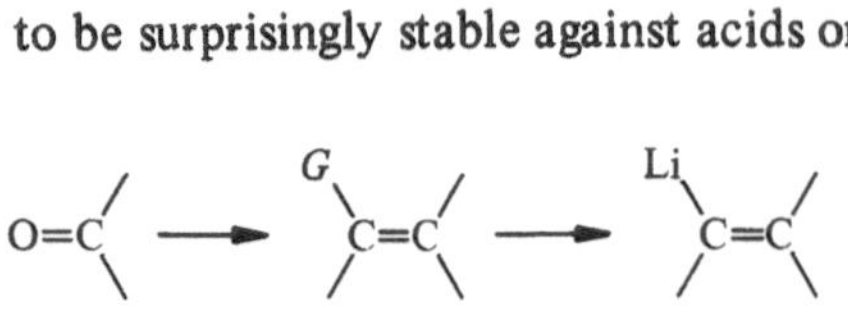

Scheme 48

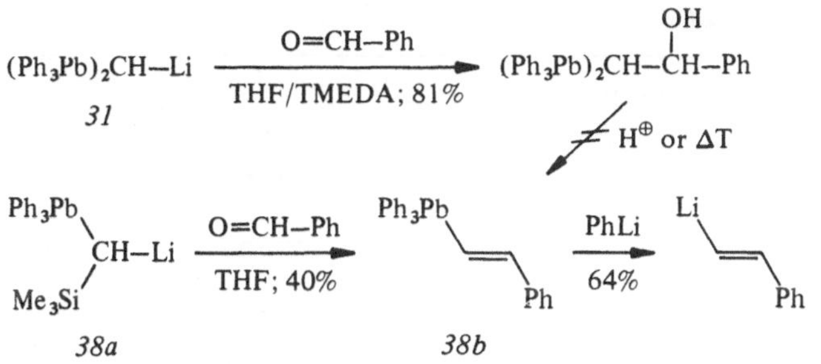

$$\text{(Ph}_3\text{Pb)}_2\text{CH--Li} \xrightarrow[\text{THF/TMEDA; 81\%}]{\text{O=CH--Ph}} \text{(Ph}_3\text{Pb)}_2\overset{\overset{\displaystyle \text{OH}}{|}}{\text{CH--CH--Ph}}$$

31

$$\downarrow\!\!\!/ \ \text{H}^{\oplus} \text{ or } \Delta\text{T}$$

$$\underset{\underset{\displaystyle \textit{38a}}{\text{Me}_3\text{Si}}}{\overset{\displaystyle \text{Ph}_3\text{Pb}}{\diagdown}}\text{CH--Li} \xrightarrow[\text{THF; 40\%}]{\text{O=CH--Ph}} \underset{\underset{\displaystyle \textit{38b}}{\text{Ph}}}{\overset{\displaystyle \text{Ph}_3\text{Pb}}{\diagdown}}\!\!\diagup \xrightarrow[64\%]{\text{PhLi}} \underset{\text{Ph}}{\overset{\text{Li}}{\diagdown}}\!\!\diagup$$

Scheme 49

stabilized by the other one and reverserly. But the combination reagent *38a* (Chap. 4) proved to be suitable for transforming carbonyl groups to triphenylplumbyl substituted olefins[46]. Since on these compounds lead/lithium exchange is possible, we have now a good method for lithioolefination of carbonyl functions. The reaction *38a* ⟶ *38b* is a Peterson type reaction[51a] because the silicium containing group is being eliminated. On account of this result we could expect that reagent *39* in a Horner type reaction[48] would lead to a corresponding tin-compound. But instead of the diphenylphosphoryl group the organotin group was eliminated at surprisingly low temperature (Scheme 50[10d]). These two examples and similar ones show clearly

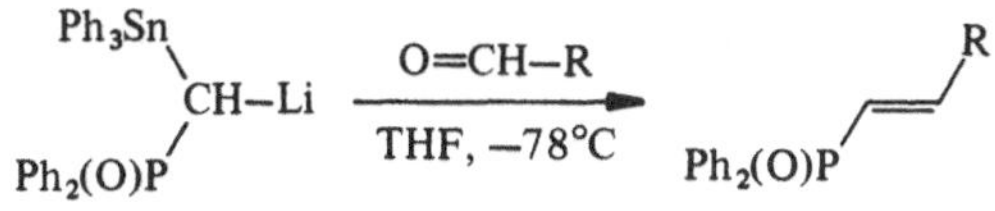

$$\underset{\underset{\displaystyle \text{Ph}_2(\text{O})\text{P}}{}}{\overset{\displaystyle \text{Ph}_3\text{Sn}}{\diagdown}}\text{CH--Li} \xrightarrow[\text{THF, }-78°\text{C}]{\text{O=CH--R}} \underset{\text{Ph}_2(\text{O})\text{P}}{}\!\!\diagup\!\!\overset{\displaystyle \text{R}}{\diagup}$$

R	%
Ph	54
Prop	53

Scheme 50

that strong mutual influences between different organoelement groups, linked together by a carbon atom, may occur. These effects — in one case increased stability of the lead-carbon bond and decreased stability of the tin-carbon bond in the other case — are by no means understood today. So we must confess not knowing enough about such combinations of organoelement groups.

6 Hydrogen/Lithium Exchange on Organometallic Compounds

The last chapter deals with hydrogen/lithium exchange. My co-worker Ennen[57, 58] reacted bis(triphenylstannyl)methane with lithiumdicyclohexylamide (LDCA) in presence of one mole HMPA and obtained the lithium compound *40* in excellent yield. This was very surprising since earlier attempts to achieve hydrogen/lithium exchange at this compounds with lithiumdiisopropylamide (LDA) had completely failed. Analogous lithiation reactions with LDA are well known on the corresponding sulfur[59], selenium[7], and tellurium[8] compound, whereas on the corresponding phophorus[60] compound lithiation was achievable with n-butyllithium (see Table 14). After the success with the tin derivative we were able to lithiate the corresponding lead, arsenic, and antimony derivative as well[58] (Scheme 51; Table 14).

Table 14. Yield of lithiation product (LDCA = lithiumdicyclohexylamide)

n-BuLi/HMPA	n-BuLi (in ether)	
$Ph_3Si–CH_2–SiPh_3$ 0%[15d, 43]	$Ph_2P–CH_2–PPh_2$ ~100%[60]	$PhS–CH_2–SPh$ ~100%[59]

LDCA/HMPA (in ether)		LDA (in THF)
$Ph_3Ge–CH_2–GePh_3$ 0%[41a]	$Ph_2As–CH_2–AsPh_2$ 63%[61]	$PhSe–CH_2–SePh$ 95%[7]
$Ph_3Sn–CH_2–SnPh_3$ 92%[57]	$Ph_2Sb–CH_2–SbPh_2$ 68%[21]	$PhTe–CH_2–TePh$ ~100%[8]
$Ph_3Pb–CH_2–PbPh_3$ 67%[10b]		

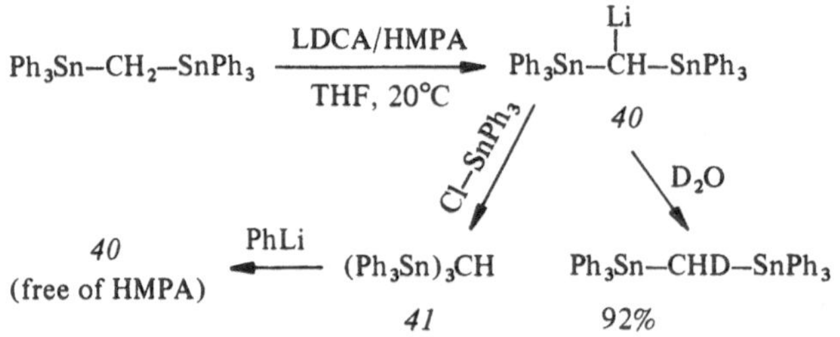

Scheme 51 (LDCA = lithiumdicyclohexylamide)

These findings confirm our result from the reported element/lithium exchange study that heavy elements are capable of stabilizing a negative charge very well.

On the contrary the triphenylsilyl and triphenylgermyl group seem to have no acidifying effect. Lithiation in this case was neither possible with lithiumdicyclohexylamide plus HMPA nor with n-butyllithium or t-butyllithium plus HMPA. In this context it is noteworthy that bis(trimethylsilyl)methane (*42a*) in contrast to bis(triphenylsilyl)methane reacts with n-butyllithium smoothly to the lithium compound *42*[62]. The striking difference is well understandable if assuming in bis(triphenylsilyl)methane the resonance formulated in Scheme *52* because this brings a

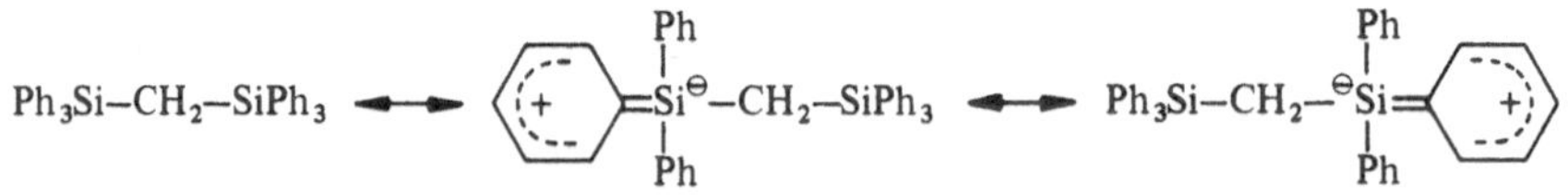

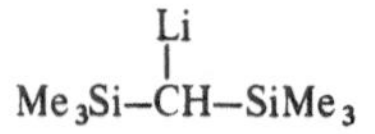

Scheme 52

partial negative charge to the silicium atoms preventing the hydrogen lithium exchange at the methylene group[63]. This explanation takes for granted a $(p \longrightarrow d) \pi$ bond between silicium atoms and the phenyl residues. The reasonable assumption of no or weak contribution of such bonds in case of the bis(triphenylstannyl)- and bis(triphenylplumbyl) — methanes would explain the significant higher acidity of these compounds very well.

Why was detected the acidifying effect of heavy organoelement groups only recently? To our mind the reason is the high electrophilicity of heavy organoelement groups. The acidifying effect is only then detectable if the attack of the base to the element can be avoided. Somewhat puzzling is the fact that by the reacting of bis(triphenylstannyl)methane with LDA/HMPA only slight lithiation to compound 40 occurs. Because LDA tends more to addition reactions than LDCA does, it is supposed that LDA forms the adduct 43 from which the starting material bis(triphenyl-

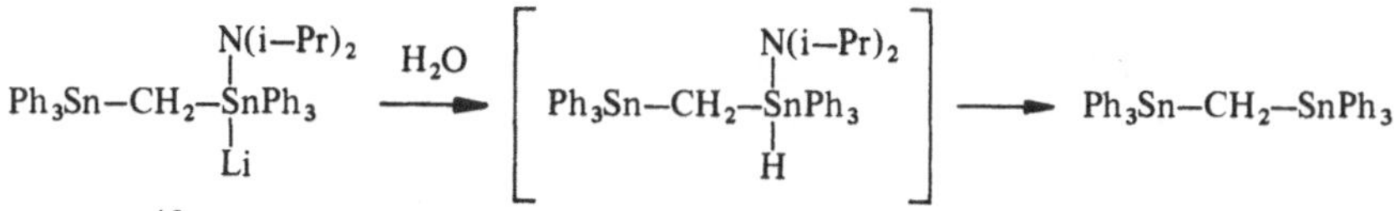

$$\underset{43}{Ph_3Sn-CH_2-\overset{\overset{\displaystyle N(i-Pr)_2}{|}}{\underset{\underset{\displaystyle Li}{|}}{Sn}}Ph_3} \quad \xrightarrow{H_2O} \quad \left[Ph_3Sn-CH_2-\overset{\overset{\displaystyle N(i-Pr)_2}{|}}{\underset{\underset{\displaystyle H}{|}}{Sn}}Ph_3 \right] \quad \longrightarrow \quad Ph_3Sn-CH_2-SnPh_3$$

stannyl)methane is regenerated by hydrolysis with water. The new lithium reagent 40 allowed us to prepare for the first time tris(triphenylstannyl)methane 41[64] (Scheme 51). From compound 41 compound 40 was obtained free of bicyclohexylamine and HMPA and therefore in a more reactive state.

7 Summary

If we look for elements which enable a C-C double bond to add organolithium compounds, *normal nucleophilic addition or cycloaddition according Scheme 53*, we are restricted to the lighter elements (Scheme 54 A) because only these are resistent

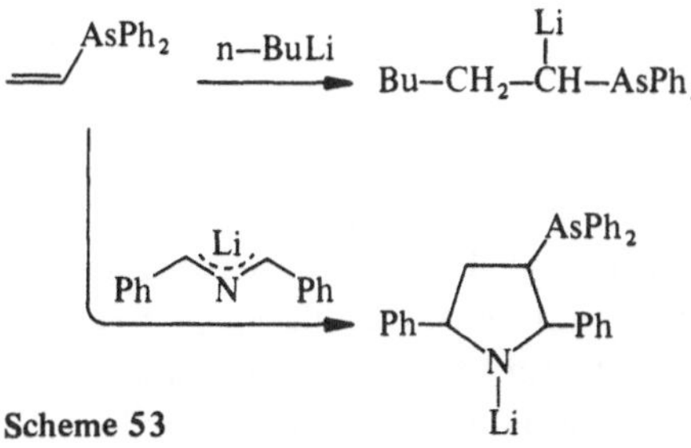

Scheme 53

to the direct attack of organolithium compounds. If an organoelement group of an heavy element is linked to a C-C double bond as in *44*, element/lithium exchange occurs instead of the nucleophilic attack to the C-C double bond. This is the explanation of the straight and clear borderline between third- and forth-row-elements in Scheme 54 A.

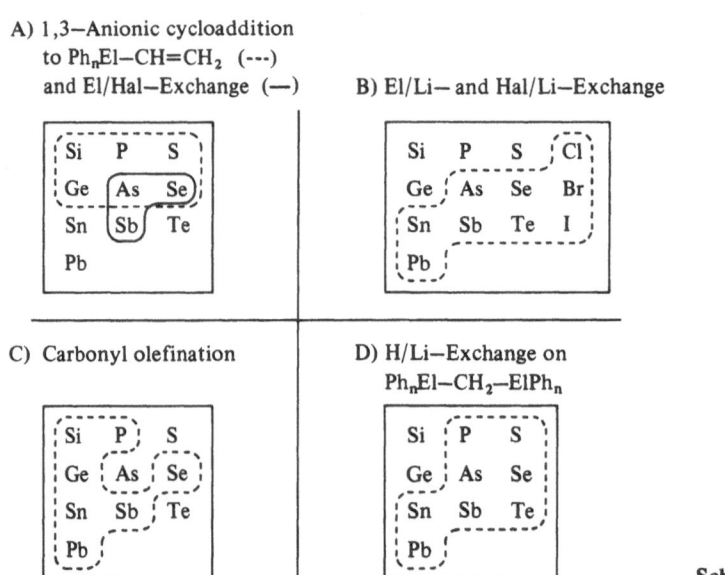

Element/halogen exchange in organic synthesis is especially advantageous with elements in the middle of the considered area of the periodic table (Scheme 54 A).

Element/lithium exchange, a reaction which has allowed us to prepare several new reagents, is especially favorable with the heavier elements (Scheme 54 B). *Carbonyl olefination* is altogether more favorable with less electronegative elements (Scheme 54 C) and the *hydrogen/lithium exchange* in compounds of the type $Ph_nEl\text{-}CH_2\text{-}ElPh_n$ is impossible only with the Si and Ge derivative (Scheme 54 D).

A) 1,3—Anionic cycloaddition
   to Ph_nEl—$CH=CH_2$ (---)
   and El/Hal—Exchange (—)   B) El/Li— and Hal/Li—Exchange

Si	P	S
Ge	As	Se
Sn	Sb	Te
Pb		

Si	P	S	Cl
Ge	As	Se	Br
Sn	Sb	Te	I
Pb			

C) Carbonyl olefination   D) H/Li—Exchange on
                                     Ph_nEl—CH_2—$ElPh_n$

Si	P	S
Ge	As	Se
Sn	Sb	Te
Pb		

Si	P	S
Ge	As	Se
Sn	Sb	Te
Pb		

Scheme 54

As it is to be seen from Scheme 54 elements in a medium position of the area investigated by us are suitable for different functions in organic synthesis. Organoelement groups of these elements can be considered as multi-purpose functional groups. For instance the diphenylstibanyl group can be exchanged for halogen and lithium, it is an acidifying group, and a good leaving group in carbonyl olefination as well.

8 References

1. New Reagents, Part 13 – Part 12: l.c. 58
2. Davidson, A. H., Hodgson, P. K. G., Howells, D., Warren, S.: Chem. Ind. (Lond.) *1975*, 455
3. a) Review: Kauffmann, Th.: Angew. Chem. *86*, 715 (1974); Angew. Chem. Int. Ed. Engl. *13*, 627 (1974). b) Full papers: Kauffmann, Th., Habersaat, K., Köppelmann, E.: Chem. Ber. *110*, 638 (1977); Kauffmann, Th., Eidenschink, R.: Chem. Ber. *110*, 645, 651 (1977); Kauffmann, Th., Berger, D., Scheerer, B., Woltermann, A.: Chem. Ber. *110*, 3034 (1977)
4. Eidenschink, R., Kauffmann, Th.: Angew. Chem. *84*, 292 (1972); Angew. Chem. Int. Ed. Engl. *11*, 292 (1972); Boche, G., Martens, D.: Angew. Chem. *84*, 768 (1972); Angew. Chem. Int. Ed. Engl. *11*, 742 (1972); Bannwarth, W., Eidenschink, R., Kauffmann, Th.: Angew. Chem. *86*, 476 (1974); Angew. Chem. Int. Ed. Engl. *13*, 468 (1974); Ford, W. T., Radue, R., Walker, J. A.: Chem. Commun. *1970*, 966; Luteri, G. F., Ford, W. T.: J. Organomet. Chem. *105*, 139 (1976)
5. Orchin, M., Jaffé, H. H.: The importance of antibonding orbitals. Boston Mass.: Houghton Mifflin 1967
6. Lehn, J. M., Wipff, G.: J. Am. Chem. Soc. *98*, 7498 (1976); Lehn, J. M., Wipff, G., Demuynck, J.: Helv. Chem. Acta *60*, 1239 (1977)
7. Seebach, D., Peleties, N.: Chem. Ber. *105*, 511 (1972)
8. Seebach, D., Beck, A. K.: Chem. Ber. *108*, 314 (1975)
9. Kauffmann, Th., Ahlers, H., Hamsen, A., Schulz, H., Tilhard, H.-J., Vahrenhorst, A.: Angew. Chem. *89*, 107 (1977); Angew. Chem. Int. Ed. Engl. *16*, 119 (1977)
10. a) Ahlers, H.: Diplomarbeit, Universität Münster 1977; b) Hamsen, A.: Universität Münster; experiments 1976–1978; c) Schulz, H.: Diplomarbeit, Universität Münster 1977; d) Tilhard, H. J.: Dissertation, Universität Münster, prospectively 1980; e) Vahrenhorst, A.: Universität Münster, experiments 1976–1978; f) Fischer, H.: Universität Münster, experiments 1976
11. Popowski, E.: Z. Chem. *14*, 360 (1974)
12. Stöckelmann, H.: Dissertation, Universität Münster, prospectively 1980
13. Busch, A.: Dissertation, Universität Münster 1974
14. a) Mozingo, R., Wolf, D. E., Harris, S. A., Folkers, K.: J. Am. Chem. Soc. *65*, 1013 (1943). b) Schönberg, A., Brosowski, K. H., Singer, E.: Chem. Ber. *95*, 2984 (1962)
15. a) Kauffmann, Th., Koch, U., Steinseifer, F., Vahrenhorst, A.: Tetrahedron Lett. *1977*, 3341; b) Koch, U.: Diplomarbeit, Universität Münster 1977; c) Steinseifer, F.: Diplomarbeit, Universität Münster 1977; d) Steinseifer, F.: Dissertation, Universität Münster, prospectively 1980
16. Kauffmann, Th., Berg, H., Köppelmann, E., Kuhlmann, D.: Chem. Ber. *110*, 2659 (1977)
17. a) Poller, R. C.: The chemistry of organotin compounds, p. 44. London: Logos Press 1970; b) Seyferth, D., Vick, S. C.: J. Organomet. Chem. *144*, 1 (1978)
18. For instance: Ashe III, A. J.: J. Am. Chem. Soc. *93*, 3293 (1971)
19. Köbrich, G.: Angew. Chem. *84*, 557 (1972); Angew. Chem. Int. Ed. Engl. *11*, 473 (1972)
20. Woltermann, A.: Universität Münster, experiments 1977/78
21. Joußen, R.: Dissertation, Universität Münster, 1979
22. Sevrin, M., Dumont, W., Hevesi, L., Krief, A.: Tetrahedron Lett. *1976*, 2647
23. Kriegesmann, R.: Dissertation, Universität Münster, prospectively 1980
24. a) Doak, G. O., Freedman, L. D.: Organometallic compounds of arsenic, antimony and bismuth, p. 201. New York, London: Wiley 1970; b) l.c. 24a, p. 325; c) Bähr, G., Pawlenko, S. in: Methoden der organischen Chemie, (Houben-Weyl-Müller), 1. Ed., Vol. XIII/6, p. 280. Stuttgart: Thieme 1978; d) Bähr, G., Langer, E., in: Methoden der organischen Chemie, (Houben-Weyl-Müller), 1. Ed., Vol. XIII/7, p. 241. Stuttgart: Georg Thieme Verlag 1975; e) Ladenburg, A.: Ber. Dtsch. Chem. Ges. *40*, 2274 (1907); Eaborn, C.: J. Chem. Soc. *1949*, 2755
25. Kauffmann, Th., Fischer, H., Woltermann, A.: Angew. Chem. *89*, 52 (1977); Angew. Chem. Int. Ed. Engl. *16*, 53 (1977)
26. Kauffmann, Th., Joußen, R., Woltermann, A.: Angew. Chem. *89*, 759 (1977); Angew. Chem. Int. Ed. Engl. *16*, 709 (1977)

T. Kauffmann

27. Seebach, D., Jones, N. R., Corey, E. J.: J. Org. Chem. *33*, 300 (1968); Seebach, D., Enders, D.: Angew. Chem. *87*, 1 (1975); Angew. Chem. Int. Ed. Engl. *14*, 15 (1975)
28. Lhotak, H.: Diplomarbeit, Universität Münster 1978
29. Sevrin, M., Denis, J. N., Krief, A.: Angew. Chem. *90*, 550 (1978); Angew. Chem. Int. Ed. Engl. *17*, 526 (1978)
30. Sinkovec, H.: Dissertation, Universität Münster, prospectively 1980
31. Altepeter, B.: Diplomarbeit, Universität Münster 1977
32. Echsler, K. J.: Diplomarbeit, Universität Münster 1978
33. Aguiar, A. M., Archibald, Th. G., Kapicak, L. A.: Tetrahedron Lett. *1967*, 4447
34. Arnold, R. T., Kulenovic, S. T.: J. Org. Chem. *43*, 3687 (1978)
35. Kauffmann, Th., Ahlers, H., Tilhard, H.-J., Woltermann, A.: Angew. Chem. *89*, 760 (1977); Angew. Chem. Int. Ed. Engl. *16*, 710 (1977)
36. Wilgen, F. J.: Diplomarbeit, Universität Münster 1978
37. Seyferth, D., Weiner, M. A.: J. Am. Chem. Soc. *83*, 3583 (1961)
38. a) Dumont, W., Krief, A.: Angew. Chem. *87*, 347 (1975); Angew. Chem. Int. Ed. Engl. *14*, 350 (1975); b) Van Ende, D., Dumont, W., Krief, A.: Angew. Chem. *87*, 709 (1975); Angew. Chem. Int. Ed. Engl. *14*, 700 (1975); c) Dumont, W., Krief, A.: Angew. Chem. *88*, 184 (1976); Angew. Chem. Int. Ed. Engl. *15*, 161 (1976)
39. Wittig, G., Maercker, A.: J. Organomet. Chem. *8*, 491 (1967)
40. Kauffmann, Th., Echsler, K.-J., Hamsen, A., Kriegesmann, R., Steinseifer, F., Vahrenhorst, A.: Tetrahedron Lett. *1978*, 4391
41. Kauffmann, Th., Kriegesmann, R., Woltermann, A.: Angew. Chem. *89*, 900 (1977); Angew. Chem. Int. Ed. Engl. *16*, 862 (1977)
42. a) König, R.: Diplomarbeit, Universität Münster 1979; b) König, R.: Dissertation, Universität Münster, prospectively 1981
43. Brook, A. G., Duff, J. M., Anderson, D. G.: Can. J. Chem. *48*, 561 (1970)
44. Kauffmann, Th., Hamsen, A., Kriegesmann, R., Vahrenhorst, A.: Tetrahedron Lett. *1978*, 4395
45. Kirmse, W.: Carbene chemistry, p. 147. New York: Acedemic Press 1964
46. Rensing, A.: Diplomarbeit, Universität Münster 1978
47. Willemsens, L. C., van der Kerk, G. J. M.: J. Organomet. Chem. *23*, 471 (1970)
48. a) Wittig, G., Geissler, G.: Liebigs Ann. Chem. *580*, 44 (1953); Wittig, G., Schöllkopf, U.: Chem. Ber. *87*, 1318 (1954). b) Horner, L., Hoffmann, H., Wippel, H. G., Klahre, G.: Chem. Ber. *92*, 2499 (1959)
49. Johnson, A. W., Lacount, R. B.: Chem. Ind. (London) *1958*, 1440; Franzen, V., Driessen, H. E.: Chem. Ber. *96*, 1881 (1963); Corey, E. J., Chaykowsky, M.: J. Am. Chem. Soc. *87*, 1345, 1353 (1965)
50. Grim, S. O., Seyferth, D.: Chem. Ind. (London) *1959*, 849; Henry, M. C., Wittig, G.: J. Am. Chem. Soc. *82*, 563 (1960); Gosney, I., Lilli, T. J., Lloyd, D.: Angew. Chem. *89*, 502 (1977); Angew. Chem. Int. Ed. Engl. *16*, 487 (1977)
51. a) Peterson, D. J.: J. Org. Chem. *33*, 780 (1968). b) Schrock, R. R.: J. Am. Chem. Soc. *98*, 5399 (1976)
52. Schlosser, M., Christmann, K. F.: Angew. Chem. *78*, 115 (1966); Angew. Chem. Int. Ed. Engl. *5*, 126 (1966); Liebigs Ann. Chem. *708*, 1 (1967); Bergelson, L. D., Barsukov, L. I., Shemyakin, M. M.: Tetrahedron *23*, 2709 (1967); Corey, E. J., Kwiatkowsky, G. T.: J. Chem. Soc. *90*, 6816 (1968)
53. Davis, D. D., Gray, C. E.: J. Organomet. Chem. *18*, P1 (1969); J. Org. Chem. *35*, 1303 (1970)
54. Kauffmann, Th., Ahlers, H., Joußen, R., Kriegesmann, R., Vahrenhorst, A., Woltermann, A.: Tetrahedron Lett. *1978*, 4439
55. Analogous observation by Seebach, D.: private communication 1978
56. Reich, H. J., Chow, F.: Chem. Commun. *1975*, 790; Rémion, J., Dumont, W., Krief, A.: Tetrahedron Lett. *1976*, 1385; Rêmion, J., Krief, A.: Tetrahedron Lett. *1976*, 3743
57. Ennen, J.: Diplomarbeit, Universität Münster 1979
58. Kauffmann, Th., Altepeter, B., Echsler, K.-J., Ennen, J., Hamsen, A., Joußen, R.: Tetrahedron Lett. *1979*, 501

59. Corey, E. J., Seebach, D.: J. Org. Chem. *31*, 4097 (1966)
60. Peterson, D. J.: J. Organomet. Chem. *8*, 199 (1967)
61. Echsler, K. J.: Dissertation, Universität Münster, prospectively 1980
62. Gröbel, B. T., Seebach, D.: Angew. Chem. *86*, 102 (1974); Angew. Chem. Int. Ed. Engl. *13*, 83 (1974)
63. For an analogous explanation of striking differences in ^{1}H-NMR spectra of compounds $C_6H_3D_2CAlk_3$ and $C_6H_3D_2CHal_3$ see Whitesides, G. M., Selgestad, J. G., Thomas, S. P., Andrews, D. W., Morrison, B. A., Panek, E. J., San Fillipo, J., Jr.: J. Organomet. Chem. *22*, 365 (1970)
64. This compound is twice described in literature (Kraus, C. A., Eatough, H.: J. Am. Chem. Soc. *55*, 5014 (1933); Mattson, D. S., Wilcsek, R. J.: J. Organomet. Chem. *57*, 231 (1973)), but in each case the product is according to the melting point or the ^{1}H-NMR-data not tris(triphenylstannyl)methane (*41*) which was obtained in our laboratory[57] by reacting of bis(triphenylstannyl)methyllithium (*40*) with triphenylstannylchloride. The melting point of *41* is 181 °C
65. Note added in proof: In the meantime we were able to realize the transformation in an olefin: A. Rensing, K.-J. Echsler, Th. Kauffmann, Tetrahedron Lett., in press

Received May 14, 1979

Orbital Correlation in the Making and Breaking of Transition Metal-Carbon Bonds

Paul S. Braterman[1]

Department of Chemistry, Glasgow University, Glasgow G12 8QQ, Scotland

Table of Contents

I Introduction

Some years ago, Dr. Cross and I put forward a description of concerted reductive elimination (and, by implication, concerted oxidative addition) processes at transition metal centres, assuming the conservation of orbital symmetry, within a single dominant configuration, for the most obvious reaction path[2]. This picture had unexpected implications which some recent work has rendered quite explicit, and which are discussed in Part II of this article.

Concerted reductive elimination can be regarded as the simplest type of pericyclic reaction at a metal. There is a clear temptation to generalise to more complicated processes, in which the metal-carbon bonds in question form part of a larger ring system. Such a generalisation would be related to the pioneering suggestions of Mango and Schachtschneider[3], as well as to more sophisticated and critical re-workings[4]. These suggestions form a special case of frontier orbital theory as was clearly realised by Fukui[5] [a], and are thus closely related to the Woodward-Hoffmann rules for reactions of purely organic system[6]. I succumb to this temptation in Part III.

Part IV of this article outlines a description of the processes involved in cleavage of a *single* metal-carbon bond, and puts forward a theory of the influences of electron count, and of configuration interaction, on metal-carbon bond stability. Finally, in Part V, I attempt a critical appraisal of the method adopted and the results obtained.

I must emphasise from the outset the limited range of this discussion. It says nothing about the effect of charge distribution, used[7] by Davies, Green and Mingos to explain regiospecificities in nuclephilic attack on coordinated hydrocarbons. It ignores the interactions of ligand orbitals through space or (I suspect more importantly) through vacant metal $(n + 1)\,s, p$ orbitals, although there is evidence[8] of this effect. No attempt is made to quantify orbital energies or degrees of overlap. Solvent effects, and steric effects, are not touched on. Last, but not least, Parts II and III are based on a simple 1-electron model, although transition metal chemistry is profoundly affected by configuration interaction, which dominates the discussion of Part IV.

II Reductive Elimination

A. General Theory

The discussion[1] by Braterman and Cross of reductive elimination from square planar or octahedral complexes is a special case of frontier orbital theory. A transition metal $L_n MR_1 R_2$ is taken to lose groups R_1, R_2 in a concerted step to give $L_n M + R_1 - R_2$. By the usual book-keeping convention, the electrons in the initial M—R σ bonds are assigned to the R groups but this is of course a mere convention of naming and does not affect the argument. The in-phase combination of metal-carbon σ bonds correlates

---

a   Fukui's detailed treatment, however, (Ref.[5], Chap. 11) follows Ref.[3] more closely than this article does

directly with the $R_1 - R_2$ σ bond. The out-of-phase combination is the HOMO (σ^*) of the $R_1^- --- R_2^-$ system. Electrons are transferred from this orbital to a suitable orbital on the metal and it is, of course, this transfer that leads one to describe the elimination as "reductive". The theory then requires that the metal should possess a vacant acceptor orbital[5) b] which is antisymmetric with respect to interchange of R_1 and R_2, and which is, or during the reaction becomes, sufficiently low in energy.

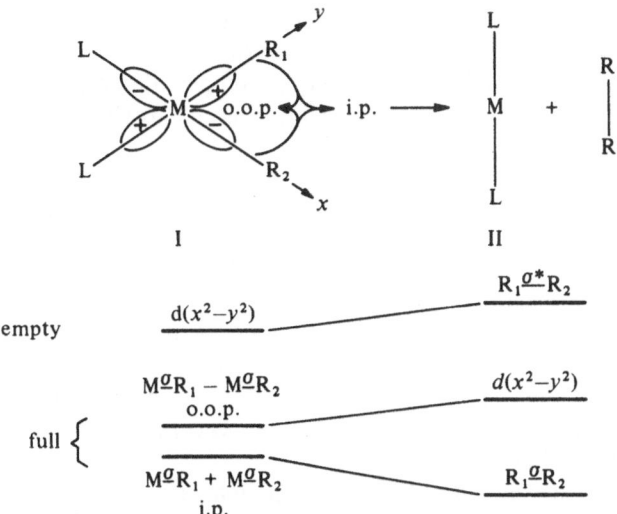

Fig. 1. Reductive elimination from a square planar d^8 metal

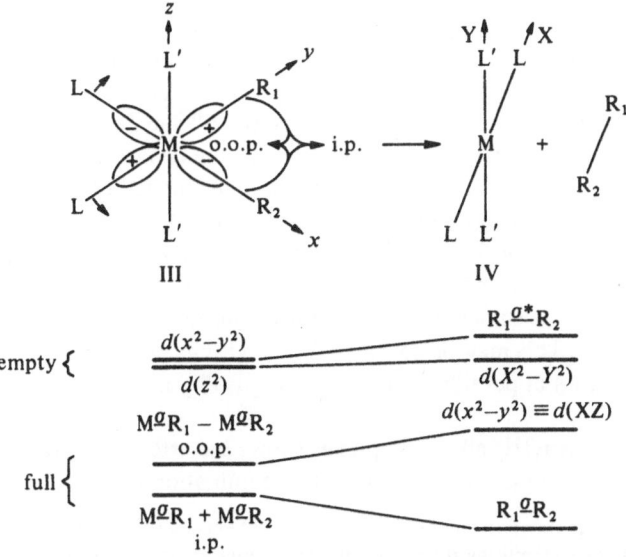

Fig. 2. "Allowed" reductive elimination from an octahedral d^6 metal with *trans*-angle opening

b   It is, of course, unimportant that this should initially be the *lowest* available unoccupied orbital

Whether this condition can be fulfilled depends on the electron count of the metal, and the stereochemistry of the elimination. For instance, in *cis*-elimination from octahedral d^6, or square planar d^8, systems, metal $nd(x^2 - y^2)$ acts as acceptor, and this should be a facile process (see Figs. 1, 2). For *trans*-elimination, on the other hand, the lowest empty orbital of correct symmetry is $(n + 1)p$. Such elimination seems energetically less likely, unless a non-concerted pathway (such as successive anionic and cationic loss) is available. The same arguments apply, of course, to oxidative additions. It follows that the many known cases of *trans* oxidative addition to square planar d^8 systems are unlikely to take place by a concerted mechanism, and this conclusion is now generally accepted[9]. There are special complexities in reductive elimination from trigonal systems, and these are discussed further in Part III.

B. Effects of Spectator Ligands

So far we have neglected the other ligands or potential ligands present. These can play several conflicting roles. For instance, they can supply electron density, so as to compensate for the removal, from the coordination sphere of the metal, of the electrons of the new $R_1 - R_2$ bond. This is presumably a partial explanation of the accelerating role of alkenes in the thermolysis of Ni(II) alkyls[10] c. Another such case is the acceleration caused by free phosphine in the thermolysis of phosphine Pt(II) aryls[13], both as melts[14] and in solution[15].

On the other hand, ligands could inhibit reductive elimination, either by destabilising the acceptor orbital, or by causing steric congestion in the transition state. There are well-established cases of both kinds. Good σ-donor ligands *trans* to the groups R_1, R_2 will raise the energy of the acceptor orbital, but this effect can be relieved in some cases by a change in ligand geometry. Where this is prevented by chelation, we should expect some inhibition in elimination from square planar systems. With octahedral systems, we should expect regiospecifity as discussed in Sect. II. D. below.

The d^{10} subshell completed by reductive elimination from a d^8 square planar system is, of course, spherically symmetric. The ligands *trans* to the departing groups will be repelled by the transferred electrons if they remain in their original place, but in just the same way they will be repelled by electrons already present, if they move apart. Nonetheless we would expect the 2-cordinate d^{10} species finally produced to prefer a linear conformation, so that monodentate phosphine complexes (such as $(Ph_3P)_2PtAr_2$) should thermolyse somewhat more readily than those of bidentate phosphines, such as dppePtAr$_2$ d. This may be so, since $(Ph_3P)_2PtPh_2$ thermolyses in solution more rapidly than does dppePtPh$_2$; but $(Ph_2MeP)_2 PtPh_2$ is also slow to react[15].

Reductive elimination from Au(III) alkyl complexes shows the effect of steric congestion. The evidence[16, 17] is compelling that reductive elimination from species

---

c However, electronegative substituents on the alkene cause greater enhancement. This suggests a synergic role for the alkene, stabilising the newly filled asymmetric metal d-orbital by its acceptor capacity in addition to its function as an electron donor. This is in effect one part of the discussion of Ref.[10] that survives[11] the abandonment[2] of the Chatt-Shaw[12] theory of bond stability

d dppe = bis(diphenylphosphino)ethane, $Ph_2PCH_2CH_2PPh_2$

LAuMe$_3$ proceeds by a dissociative pathway, while elimination from AuMe$_4^-$ is slower than that from AuMe$_3$ itself. It is not at all obvious why this should be so, but extended Hückel calculations show energetically unfavourable interactions developing during the dissociation process between the spectator ligands and hyperconjugative orbitals on the departing groups. This effect should be greater for saturated than for unsaturated leaving groups, and may explain Young's observation[15] that reductive elimination of sp^3 carbon from Pt(II) occurs, if at all, with great difficulty, while loss of sp^2 carbon is well characterised.

C. Effect of Electronic Configuration

One would naively expect the ease of reductive elimination to fall from d^8 through d^7 to d^6 systems, and there is some evidence that this is so. For instance, the nickel-catalysed Grignard cross-coupling reaction proceeds too rapidly for a cycle involving reductive elimination from Ni(II), and a cycle based on Ni(I) − Ni(III) interconversion is preferred[18]. Pt(II) d^8 metallacyclopentanes decompose by β-hydride elimination, even though this is strongly discouraged by the geometry of the ring. When Pt(IV) species are obtained from these by oxidative addition, reductive elimination becomes the preferred decomposition mode[19].

D. Ligand Chelation and Regiospecific Elimination

Species III formed by *trans*-addition of e.g. ClCH$_2$−Cl to Pt(II) d^8 metallacyclo-pentanes show the effect of ligand constraint on the selectivity of reductive elimination. Complexes in which L is a monodentate phosphine decompose preferentially

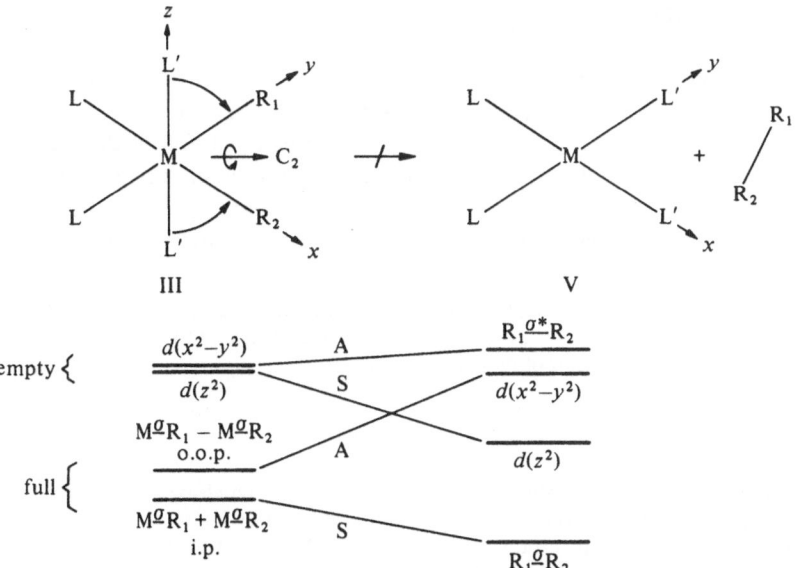

Fig. 3. "Forbidden" reductive elimination from an octahedral d^6 metal with *cis*-migration

153

by reductive elimination of cyclobutane, even though this is a strained species, and this is presumably an example of the *trans*-effect of coordinated P. However, complexes of bidentate ligands (bipyridyl, bidentate phosphine) do not show this reaction, but give rise to products from hydride shifts (open C_4 species) or from reductive elimination between ring and non-ring carbon (C_5 species). This change in behaviour can be related to the prevention of LPtL angle opening[20]. Reductive elimination, accompanied by an opening of the angle between the ligands *trans* to the departing group, is a symmetry-allowed process and minimises repulsion between those ligands and the electrons transferred (Fig. 2). Elimination without opening of the angle would still be symmetry-allowed, but would give rise to a far larger ligand field destabilisation. Elimination with concomitant migration of the ligands initially *cis* to both departing groups is a "forbidden" process on the one-electron model using idealised (C_2) geometry (Fig. 3). There will be some lowering of activation energy due to breakdown of the one-electron approximation, but in this case the effect can be shown to be small[e].

There is an important difference between the role proposed above for the supporting ligands, and Mango and Schachtschneider's distinction[22] between "restrictive" and "non-restrictive" ligand fields. Their view was of ligands static throughout a reaction, but prevented by symmetry, in certain favourable cases, from interfering with the process. The view put forward here is that supporting ligands will move, if they can, in such a way as to lower the energy of the reacting system, and that such motion is in fact a component of the reaction coordinate itself.

III Coupling Reactions at Metal Centres

A. Some General Remarks

Some of the colleagues with whom I discussed my intention to write on this topic expressed polite scepticism. There is good historical reason for this. Several previous writers had amid general admiration used frontier orbital theories to rationalise such processes as concerted *trans*-addition, square four-carbon arrays in olefin metathesis, chelation of a metal by two strained carbon-carbon σ-bonds, and trigonal bipyramidal Fe(II) alkyls. Belief in these is less widespread than it was five or ten years ago, and

---

e Inspection of Griffiths' tables[21] shows that the pure l-electron configurations $(xy)^2 (yz)^2$ $(zx)^2 (z^2)^2$ and $(xy)^2 (yz)^2 (zx)^2 (x^2 - y^2)^2$ (S^2 and A^2 in the notation of Fig. 1) correspond to linear combinations of terms of d^8 in O_h:

$$S^2 = ({}^1A_1 e_g^2 - {}^1E(\theta) e_g^2)$$
$$A^2 = ({}^1A_1 e_g^2 + {}^1E(\theta) e_g^2)$$

This classification in O_h suffices for discussion of electron-electron repulsion, because the two terms differ in the assignment of *two* electrons, and thus can only interact through the (d-) electron-electron repulsion operator, which is independent of ligand geometry.
The matrix element linking the two configurations is $4B + C$, which for Pd(II) is around 6000 cm^{-1}, which is far smaller than the ligand field splittings involved

there is a healthy suspicion of arguments that fail to discriminate between what does in fact occur and what does not.

My own credentials are mixed. On the credit side is a refusal[2] to predict concerted *trans*-addition; less impressive is a clear explanation[23] of facile free radical loss from titanium tetramethyl complexes, shortly before evidence began to accumulate[24] that such species decompose by internal or mutual abstraction.

The purpose of this section is to apply symmetry arguments to the important reaction sequences in which metallacycles are involved, while refraining from postulating novel bond types or coordination geometries. Such things do no doubt remain to be discovered, but should not[25] be introduced into the discussion without good reason.

B. The Coupling and Uncoupling of Alkenes

There are a large number of reactions of unsaturated organic ligands at metal centres which involve formal electron pair transfer from the metal to the ligand system, and the resultant formation of a new σ-bond between the ligands. Such reactions can occur with very high stereo- and regioselectivity, as in the coupling of alkenes by iron carbonyls to give cyclopentanones. A dramatic example of such selectivity is the

VI   VII

coupling of norbornene-4-one(VI) to give specifically the *syn-exo-trans-exo-syn* product VII[26]. This process represents five stereochemical choices, including, with racemic VI, a preference for coupling between molecules of the same chirality. Closely related is the photoinduced coupling of acrylic ester by iron carbonyl to give cyclopentanone- 1,5-dicarboxylic esters XI[27]. The effect of light is to remove CO from the metal at low enough temperatures for intermediates to be detected, and the reaction is known to proceed according to Scheme 1. Coupling takes place between unsubstituted rather than substituted carbon atoms. This selectivity can be understood[28] if we regard the reaction as a concerted reductive coupling dominated by frontier orbital effects. The new σ bond correlates with an in-phase combination of π^* orbitals on the parent alkenes. But in these alkenes, π^* is concentrated on the unsubstituted carbon atom[29], and so it is between these carbon atoms that the new bond forms.

At first sight, reductive coupling in VIII to give IX looks like a 6-electron pericyclic process, and therefore allowed; it also looks like a ligand field destabilised process with regard to the equatorial CO group. The reality is more complex. If charge transfer from metal to alkenes were to proceed in VIII in its ground state geometry, the energy minimum in which the equatorial CO group sits would become

155

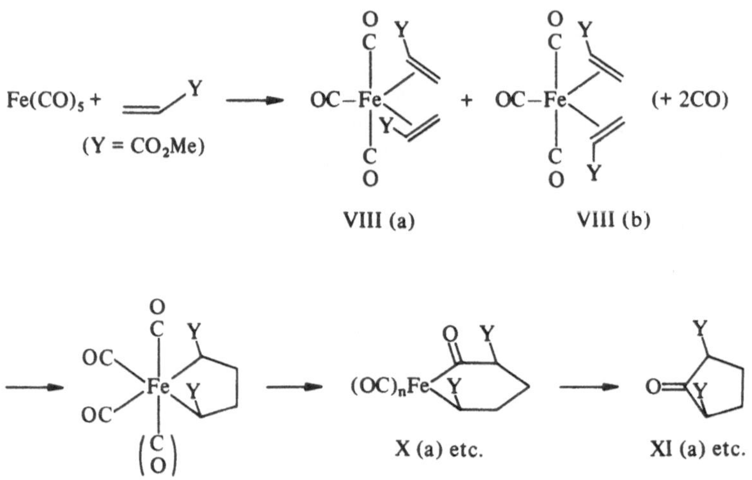

VIII (a)   VIII (b)

IX (a) etc.   X (a) etc.   XI (a) etc.

Scheme 1

progressively shallower, and would eventually be transformed into a maximum. Presumably the CO would move in the mirror plane of the complex as soon as it was energetically preferable to do so, and would take up a position *trans* to one of the new Fe-C σ bonds to give a square pyramidal low-spin Fe(II) complex. The entire process might well be facilitated by an incoming ligand, such as CO, alkene, or solvent. Unfortunately, this plausible-seeming process is symmetry forbidden. For (as Pearson[30] pointed out some years ago) simple arguments show that the electron pair transferred should come from an S orbital ($\pi(1-2) \pm \pi(3-4)$ span S + A, as do σ(Fe-1) $\pm \sigma$(Fe-3); $\sigma(2-4)$ in the product is of type S and should therefore correlate with an S orbital in the reagent). The problem then is to find a suitable S orbital, bearing in mind the fact[31, 74] that ligand fields in metal carbonyls and alkyls are large.

The product of reductive coupling (assisted or unassisted by an incoming ligand) is an octahedral or square pyramidal low-spin Fe(II) complex, in which two d-orbitals are unoccupied. Of these, one is S and the other is A with respect to interchange of the two Fe-C σ-bonds. But an S orbital has been emptied in converting the d^6 product. It follows in that d^8 precursor, the empty d-orbital was of type A. This condition is met, not by VIII, but by the readily accessible isomers XII or XIII. The argument is summarised in Table 1. The reductive coupling is indeed allowed (as it must be if the theory is to be of any value in this context), but only if it is preceded by an isomerisation. Such isomerisation is well accepted as a facile process, for reasons quite unconnected with this work[32].

The following general points emerge from our lengthy discussion: —

i) It will not do to predict the allowedness of pericyclic reactions at a metal unless account is taken of any discontinuities introduced by the metal orbitals involved.

ii) Facile isomerisations can sometimes affect allowedness,

iii) The selection rules will in any case depend on the detailed configuration at the metal.

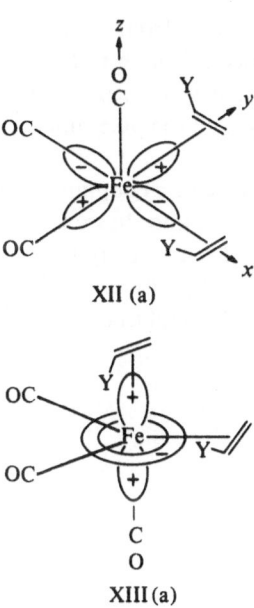

XII (a)

XIII (a)

Table 1. MOs of a metallacycle $(OC)_n Fe(CHYCH_2CH_2CHY)$ (IX) and possible precursors VIII, XII, XIII, classified under interchange of $1-2$ with $3-4$ [a]

IX	Full:	$\sigma(M-1) \pm \sigma(M-3)$	S, A	
		$\sigma(2-4)$	S	4S, 2A
		xy, yz, zx	S, S, A	
	Empty:	z^2	S	
		$x^2 - y^2$	A	
		$\sigma^*(2-4)$	A	
VIII	Full:	$\pi(1-2) \pm \pi(3-4)$	S, A	
		yz, zx	S, A	3S, 3A
		$xy, x^2 - y^2$	S, A	
	Empty:	z^2	S	
		$\pi^*(1-2) \pm \pi^*(3-4)$	S, A	
XII	Full:	$\pi(1-2) \pm \pi(3-4)$	S, A	
		yz, zx	S, A	
		xy	S	4S, 2A
		z^2	S	
	Empty:	$x^2 - y^2$	A	
		$\pi^*(1-2) \pm \pi^*(3-4)$	S, A	
XIII	Full:	$\pi(1-2) \pm \pi(3-4)$	S, A	
		xy, yz^*	S, A	
		zx	S	4S, 2A
		$x^2 - y^2$	S	
		z^2	A	
		$\pi^*(1-2) \pm \pi^*(3-4)$	S, A	

[a] Interconverted degenerate or near-degenerate pairs span S, A

Thus in our example, the only electron pair available for transfer from the metal was initially in an antisymmetric orbital, and this made the process symmetry-forbidden; isomers were accessible in which the donor orbital became symmetric, thus allowing the reaction; and the specific electron pair availability, and the possibility of facile isomerisation, are both characteristic of a d^8 starting state.

If the above arguments are correct, it is possible for the course of a coupling reaction at a metal to depend on the number and nature of the spectator ligands. This is so, as shown by Grubbs' reports of the reactions of metallacycles of type $L_n Ni(CH_2)_4$[33, 34]. It is reported that the preferred decomposition pathway depends on the degree of coordination (Scheme 2). Here the formation of cyclobutane from

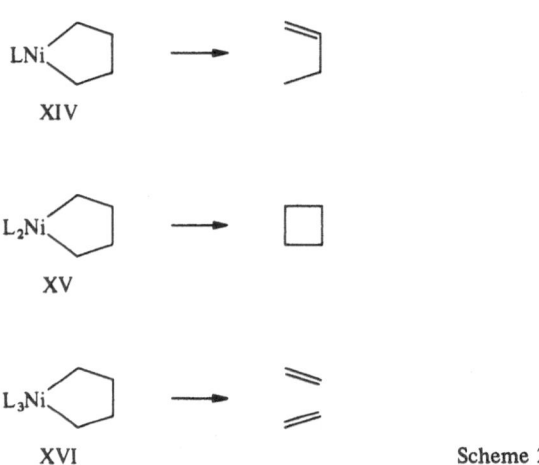

XIV

XV

L$_3$Ni

XVI   Scheme 2

XV is presumably a straightforward reductive elimination from a square planar d^8 species, and allowed (that Ni(II), unlike Pt(II), readily undergoes this process even with sp^3 carbon ligands is clear from work already cited[10]). Less saturated species (XIV) decompose, presumably by β-elimination followed by reductive elimination, to give 1-butene. More surprisingly, the increase in coordination number in XVI leads to the formation of coordinated ethylene, i.e. to the formal reverse reaction to the coupling at iron discussed above. This is a strange-seeming result, since a *higher* coordination number leads to the formation of *more* potential ligands, but it is a consequence[20] of orbital symmetry requirements. For the decoupling of two ethylene fragments requires the transfer to the metal of an electron pair from the C(β)-C(β') σ-bond, which is of course symmetrical. Such transfer is symmetry-forbidden in XV, in which the only vacant metal d-orbital is $d(x^2 - y^2)$ (antisymmetric). However, in the (e, e) trigonal bipyramidal isomer of XVI, the potential acceptor orbital is $d(z^2)$, and the reaction becomes allowed. Presumably the uncoupling of the C_2H_4 fragments is accompanied by the departure of the other equatorial ligand, as in Fig. 4. If so, then we are witnessing a new type of nucleophilic displacement at the metal,

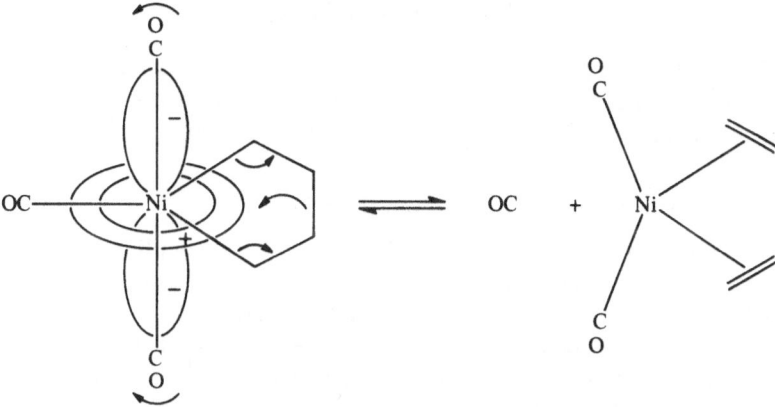

Fig. 4. "Allowed" decoupling reaction in $L_3Ni(CH_2)_4$

in which an electron pair transferred from beyond the initial coordination sphere expels the lone pair of one of the original ligands[f].

The oxidative decoupling reaction of the tetramethylene ligand should be allowed whenever the metal carries a vacant low-lying symmetric orbital. Thus the reaction is shown at low temperature by $(\eta^5\text{-}C_5H_5)_2\,Ti(CH_2)_4{}^{35)}$, but XVII, in which the only completely vacant d-orbital on each metal is antisymmetric towards each $(CH_2)_4$

XVII	XVIII

L = PhMe$_2$P

group, is thermally more stable[36]. It may be relevant that XVIII shows the uncoupling reaction photochemically but not thermally[37], but there is more than one possible explanation for this. Uncoupling in a metallacyclobutane will give a metal (carbene) (alkene) complex. Such processes are now generally thought to be relevant to alkene metathesis[38], although the low symmetry of the process makes difficult the correlation of reagent and product orbitals.

---

f   An alternative view (R. H. Grubbs, private communication; compare also Ref.[34]) is that the decoupling reaction occurs in tetrahedral or otherwise distorted $L_2Ni(CH_2)_4$. In this case we might be witnessing the breakdown of selection rules by configuration interaction, on close approach of $d(x^2 - y^2)$ and $d(z^2)$, or indeed a totally different reaction pathway involving intersystem crossing from a triplet ground state

C. Relationship of Present Work to Woodward-Hoffmann Rules

For more than a decade, thermal pericyclic reactions in organic chemistry have been classified as "allowed" or "forbidden" according to the number of electron pairs, and of phase reversals, involved in the reaction[6]. In general, a reaction is thermally "allowed" if

$$(\text{number of electron pairs}) + (\text{number of phase reversals}) = 2n + 1 \quad {}^{g} \qquad (1)$$

where n is an integer. Phase reversals occur as a result of antarafacial coupling or of its converse, conrotatory ring opening. The rules are closely related to those for aromaticity. They have much generality and can be derived in various ways, one of the simplest being to consider the relative phases at the extremities of the $2n$'th and $(2n + 1)$ th wave-function of an electron on a wire. Individual examples are commonly discussed by reference to the "frontier" (highest filled and lowest empty) orbitals[5] of interacting fragments.

Pericyclic reactions of a metal present some new features. The energy gaps between orbitals may be quite small, and so (as Fukui[5] points out) there seems no reason why the orbitals involved in a reaction should be the actual highest filled or lowest empty ones. The energetics of a reaction depend on the electrons and orbitals actually involved, and orbitals of more suitable energy but unsuitable phase properties would then be irrelevant rather than inhibiting. In any case, the concept of "energy" is less well defined for a metal- than for an carbon-centred orbital. The number and spatial arrangement of ligands can and sometimes must change during a reaction; the oxidation state and electronegativity will also probably change; and orbital relaxation effects on electron transfer are more serious[39] for transition metals than for carbon.

For these reasons, Eq. (1) must be applied with care. It will not do to neglect the possibility of a phase reversal at the metal. It seems far better to use Eq. (1) to predict whether any such reversal is required. The next step is to see whether energetically suitable orbitals of the correct symmetry and occupancy do exist in the actual complex being considered, and, if not, how they can be brought into being. We can then advance from considering the allowedness of a reaction to discussing its probable detailed course.

A very special complication can arise from the breakdown of the single configuration approximation. This is most likely to be a problem in systems where the d-orbital energy ordering alters drastically during a reaction; an example may be the butadiene coupling reaction discussed below.

D. Reductive Coupling of Butadienes

This reaction has been extensively studied[40−42] and I discuss it here merely to point out certain pitfalls. The conversion of a Ni(O) bis (butadiene) complex XIX to a Ni(II) bis- π-allylic complex XX (the structures are drawn without stereochemical implication) by reductive coupling appears at first sight to be an allowed process; each ligand contributes four electrons and the metal contributes two, making ten in all. Thus Eq. (1) is satisfied, provided the metal itself introduces no discontinuity[30].

---

g   This is, of course, equivalent to the much-quoted formulation of Ref.[6]

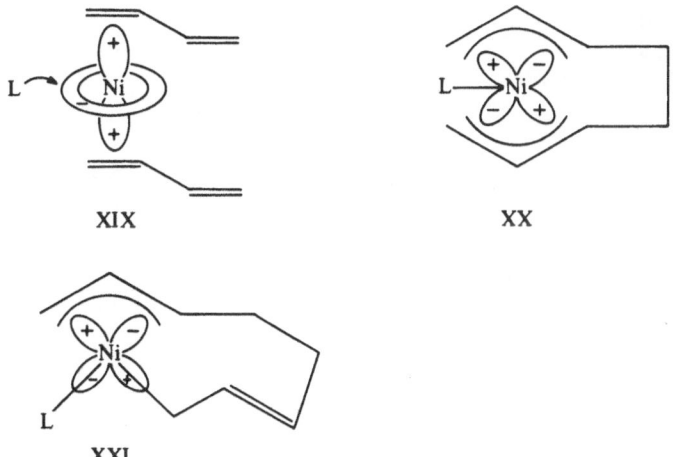

XIX   XX

XXI

Unfortunately, the orbital emptied does just that. The difficulty can be avoided by re-formulating the coupling product as XXI, a suggestion made some years ago for other reasons[41]. There remains the serious possibility that the single configuration approximation breaks down. For if, as seems possible, $d(z^2)$ is the highest (filled) d-orbital in XIX but $d(xy)$ is the highest (and only empty) d-orbital of XX, the configurations d^8 [$d(z^2)$ empty] and d^8 [$d(xz)$ empty] would not cross but mix[h] and the simple use of correlation diagrams would be inappropriate.

E. Rearrangements of Strained Hydrocarbons

There is an important class of rearrangements of strained cyclic σ-bonded systems to give less strained π-bonded systems which occur under the influence of transition metal catalysts although the uncatalysed process is Woodward-Hoffman forbidden and slow. Examples are the conversion of cubanes XXII and bis-homocubanes XXIII to *syn*-tricyclooctadienes XXIV and related species XXV[4, 43, 44] and of quadricyclene (XXVI) to norbornadiene (XXVII)[45]. [Ag$^+$, however, converted cubane and related species to the previously unrecognised species cuneane (XXVIII) and its relatives[44, 46–48], as do some electrophiles with incompletely filled d-subshells[44, 46]. Such processes are presumably related to the large class of Ag$^+$-catalysed rearrangements of bicyclobutanes[49]. These proceed by highly asymmetric steps, and lie outside the scope of this review]. The rearrangements we are discussing can proceed along facile, symmetry-allowed pathways according to Scheme 3[14]; using [Rh(CO)$_2$Cl]$_2$ the intermediate metallacycle can be trapped[43] by subsequent CO insertion, giving XXIX. Step (*a*) of Scheme 3 is an oxidative insertion, the reverse of the reductive elimination discussed in Sec. II. Step (*b*) is a metallacycle uncoupling, analogous to the uncoupling in L$_3$Ni(CH$_2$)$_4$ and η^5-C$_5$H$_5$)$_2$Ti(CH$_2$)$_4$, and the reverse of the ferracycle formation, all discussed above. Thus no new reaction types

---

h The matrix element in this case would be B + C, or around 4500 cm^{-1} for a Ni(II) complex

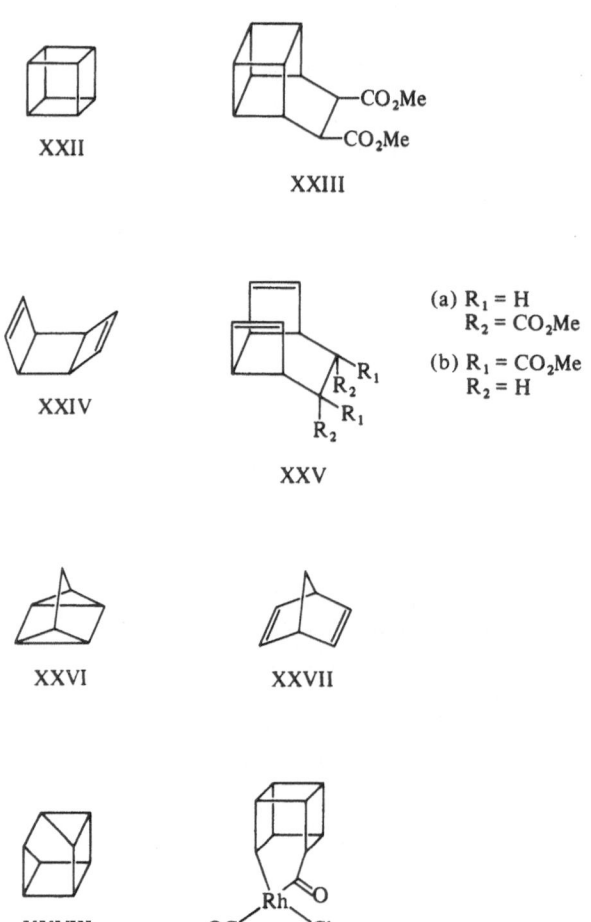

XXII

XXIII

XXIV

(a) $R_1 = H$
     $R_2 = CO_2Me$

(b) $R_1 = CO_2Me$
     $R_2 = H$

XXV

XXVI

XXVII

XXVIII

XXIX

need be invoked. As Mango[50] has pointed out, the two-step scheme retains one significant feature of the original Mango-Schachtschneider one-step scheme[3]; the otherwise symmetry-forbidden rearrangement of a cyclobutane to a pair of olefinic fragments is systematically accompanied by a change in the symmetry of the vacant orbital on M from S to A with respect to interchange of those fragments.

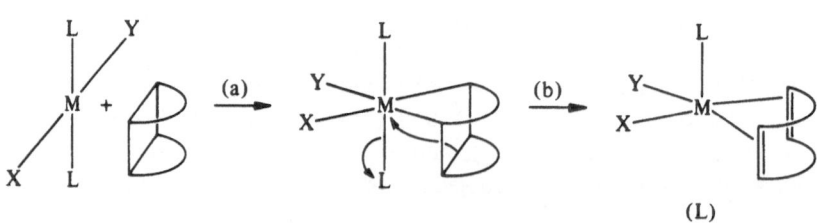

Scheme 3

The formal reverse of this process is the dimerisation of XXX to XXXI[51]. This process is catalysed by Pd(II) halides and by (η^3-allyl PdCl$_2$), but not by (Ph$_3$P)$_2$-PdCl$_2$, thus illustrating the importance of coordinative unsaturation in the catalyst.

XXX 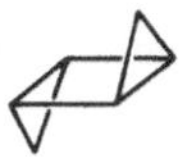   XXXI

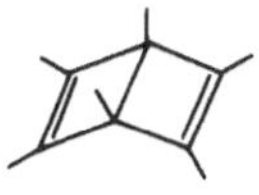

XXXII 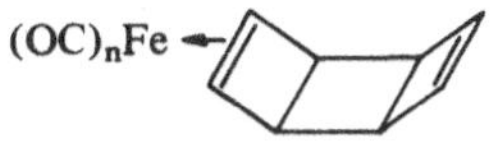   XXXIII (a) n = 4
                      (b) n = 3

XXXIV

Unsaturation is also important in the metal-catalysed disrotatory ring opening[52] of XXXII ("hexamethyl-Dewar-benzene", HMDB) to hexamethylbenzene. This formally forbidden process is catalysed by monomeric HMDBRhCl (the reaction being of order 1/2 in [HMDBRhCl]$_2$ and order 1 in substrate)[53]. Closely related is the conversion of XXXIII(a) to XXXIV, presumably by way of an unsaturated intermediate XXXIII(b), since free CO or added alkene ligands inhibit the process[54]. A plausible intimate mechanism is that of Scheme 4, in which an A electron pair on the metal and the S electron pair of the cleaved metal-metal bond correlate with out-of-phase and in-phase combinations of metal-carbon σ-bonds. The difference between XXXVI(a) and the more usual formulation XXXVI(b) is merely apparent. An electron pair formally assigned in the former to Ψ_3 of the conjugated C$_4$ ligand (the metal being d^6) is equally formally assigned in the latter to a metal being d^8). Scheme 3 is of course closely related to the scheme of Mango and Schachtschneider's seminal paper[3], but their requirement that the reaction involve coordination of the σ-bond to be cleaved has here been dropped.

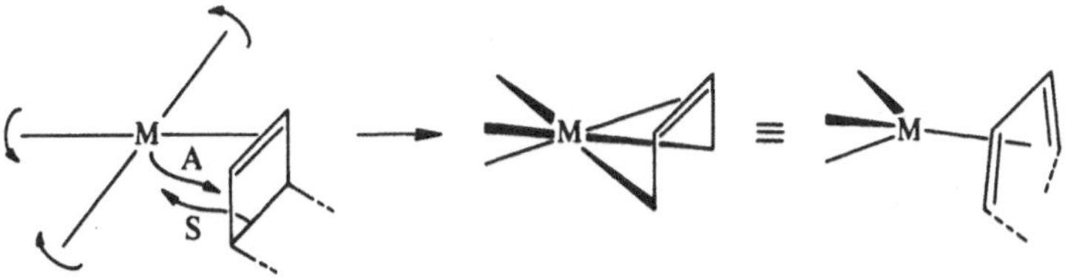

XXXV   XXXVI (a)   XXXVI (b)   Scheme 4

F. Some Examples from Acetylene Chemistry

The reactions of acetylenes and of unsaturated metallacycles present a large and difficult field[55]. The high reactivity of the species involved gives rise to a rich and di-
verse chemistry but this does not make mechanistic interpretation any easier. As is
well known[56], the bis(acetylene) — cyclobutadiene interconversion is difficult to
rationalise, but nonetheless it does take place[57].

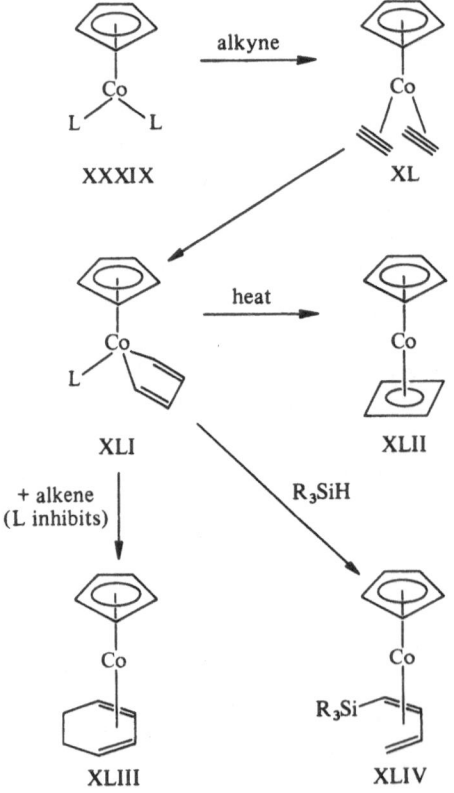

XXXVII   XXXVIII

Some light is shed of these processes by recent studies of the reactions of cyclo-
pentadienyl-cobalt(I) species with acetylenes. For example, XXXVII reacts with
diphenylacetylene to give XXXVIII, but relatively high temperatures (around 120 °C)
are necessary. Some of the intermediates involved have now been isolated and their
reactions studied (Scheme 5)[57-60]. The conversion of XL to XLIVis a typical sym-

XXXIX   alkyne   XL

XLI   heat   XLII

+ alkene
(L inhibits)   R₃SiH

XLIII   XLIV   Scheme 5

metry-allowed reductive coupling; the Co(I) of XL has only one vacant d-orbital, of A symmetry, while Co(III) in XLI has one A and one S d-orbital vacant; the electron pair lost thus correlates with the new carbon-carbon σ-bond. XLI can be converted to a butadiene-type complex either by an alkene or by a silicon hydride. Both these processes are symmetry-allowed; the former is related to the Diels-Alder reaction while the latter is a form of 1,4 addition. The "forbidden" rearrangement of XLI to XLII requires generally higher temperatures and it would be interesting to know more about the mechanism of this reaction, and in particular whether it is inhibited by added ligands, whether it is unimolecular or bimolecular in metal[56], and whether highly asymmetric intermediates, such as XLV, play any role.

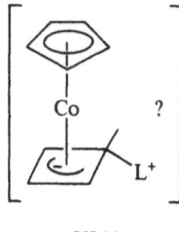

XLV

IV The Cleavage of Individual Metal-Carbon Bonds

A. General Theory

The classical view was that σ-bonds between transition metals and carbon were generally weak, and that this weakness was connected with the possibility of electronic rearrangement during a bond rupture process[12, 61]. Most recent authors have preferred the view that this is not so, but that such bonds are frequently labile for mechanistic reasons[i 62]. Some recent attempts have been made[23, 63] to relate bond lability to electronic configuration at the metal, but these are open to criticism[j], and it is time to attempt a more coherent treatment.

When we try to do this, we quickly find that the predicted patterns of kinetic stability depend on several factors:
a) the assumed coordination geometry;
b) the assumed strength of the fields of the other ligands at the metal;
c) the assumed *fate* of the electrons originally in the metal-carbon σ-bond;
d) the assumed degree of repulsion (in the starting material) between those electrons in the metal-carbon σ-bond and those formally assigned to metal d-orbitals.

---

i Ref.[12] carefully avoids specifying whether the rupture was to be homolytic or heterolytic. This, the authors have assured me, was deliberate

j I have criticised Ref.[23] (my own work) in Sect. III. A. above; Ref.[63] I criticised at the time[64]

In this article, for clarity, I shall restrict myself to initial octahedral coordination. I shall consider loss of only one σ-bonded ligand, which I designate as lying on the z-axis. I represent the initial metal-carbon bond as a σ-bond between some orbital on carbon and the $d(z^2)$ orbital of the metal. This is a less damaging simplification than it may seem. A better orbital to choose would in principle be some kind of nd, $(n + 1)s,p$ hybrid. This would powerfully concentrate the valence electron on one side of the metal, but would have a much less marked effect on the repulsion energy between that valence electron and the other d-electrons, since the electron densities of the latter are symmetrical under reflection in the xy plane. The spin of the valence electron is indeterminate, and in the absence of other clear arguments I assign statistical weights to the multiplicities arising from the two possible values. More detailed calculations[65] show that this weighting makes little difference.

The "valence state' of a formally d^n metal is then easily found, in the strong field formalism, using Griffith's coupling constants[21]. The d^n manifold can be assumed to be either high spin or low spin. The n electrons are then distributed over $t_{2g} + e_g(x^2 - y^2)$, or t_{2g} only, depending on the assumptions made, and the required terms of the d^{n+1} manifold are found by inspection. If the t_{2g} subshell contains 1,2,4, or 5 electrons then two solutions are possible; that of lower energy is chosen. Obviously, the theory in this form cannot be applied for n > 8.

I can now assign formal fates to the electrons "in" the σ-bond to be broken. The cleavage can proceed by R^- loss, $R\cdot$ loss, or R^+ loss, to give (before electronic reorganisation is considered) d^n, $d^n d(z^2)$, or $d^n d(z^2)^2$ configurations.

I must emphasise that this classification is formal rather than mechanistic and that I do *not* wish in any way to imply a unimolecular unassisted bond-breaking. My choice of reaction coordinate implies a neglect of some kinds of assistance at the metal, but the reaction at carbon can be assisted in any of several different ways. For instance, if the metal is lost by $S_N 2$ attack on coordinated carbon, this constitutes R^+ loss, and alkyl migration to an electrophilic centre such as coordinated CO may resemble R^- loss. $R\cdot$ loss may take place by simple homolysis, or[66] by alkyl group transfer. Moreover, as Yamamoto has pointed out[67], an electroneutral metal-carbon bond lengthening may be a prelude to more complex processes such as β-elimination, or may lead to internal hydrogen abstraction rather than to actual free ligand release.

B. Results for Octahedral Complexes

The treatment of loss of R^- from d^0 systems is trivial; dissociation occurs along some kind of bond dissociation curve to give, in the imagined limit, d^0 metal and a carbanion. In solution at least, both fragments will interact with the environment at bond distortion energies well below the dissociative limit.

$R\cdot$ loss is more interesting[23] (Fig. 5). The model valence state of the metal is $^2Eg\,(e_g^1)$. Where the bonding M-R interaction has fallen until it is equal in magnitude to the sum of the remaining antibonding ML interactions, this state crosses that state $^2T_{2g}(t_{2g}^1)$ which of course becomes the ground state in the dissociated limit. (There is, of course, the interesting possibility that $R\cdot$ will depart at an angle to the

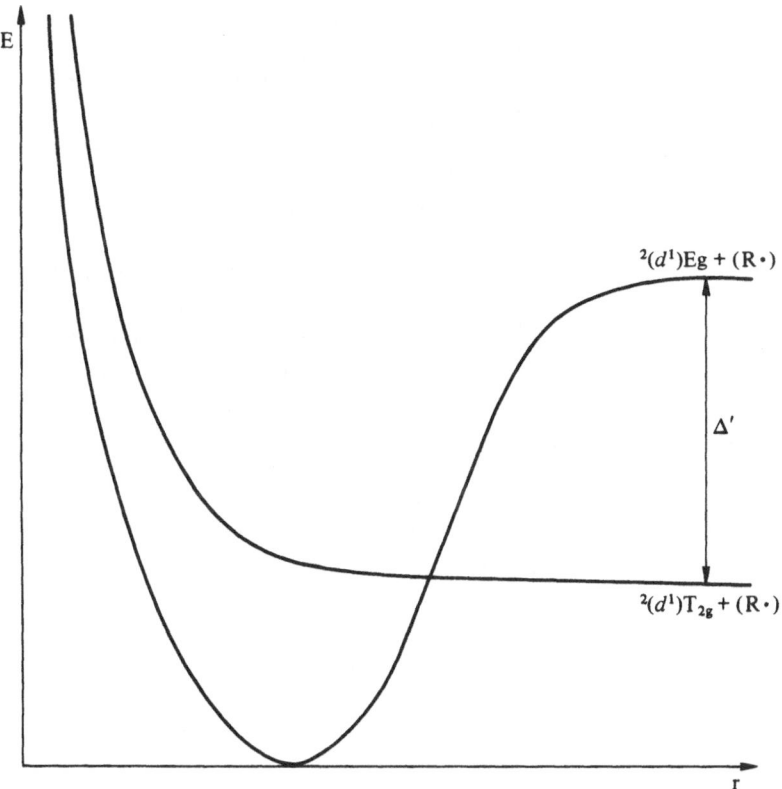

Fig. 5. Electroneutral ligand loss from an octahedral d^0 metal

z-axis, thus mixing $d(z^2)$ with $d(xz)$ and/or $d(yz)$. This is an example of more general phenomena — that the Jahn-Teller theorem applies to transition states[68], and that if a symmetrical pathway for a transition involves a change in orbital symmetry, a less symmetrical path is preferred[69]. We must ignore such complexities for the moment.) Thus in the imagined limit, the energy required for bond rupture is given by the binding energy in the M-R bonding orbital, less the promotion energy to the valence state from the ground state. This promotion energy is the ligand field energy of $d(z^2)$ in square pyramidal ML_5, which in the simplest theory[70] is equal to $2/3\ \Delta_0(ML_6)$. Free radical loss is thus facilitated by the existence of a vacancy in the t_{2g} subshell; this is merely to say that such loss is a reduction. To claim this as compelling reason to expect free radical loss from d^0 systems, as I at one time did[23], is to go beyond the evidence and such presumption was soon punished[24].

Intellectually the most interesting case for a d^0 system is loss of R^+. This might be thought to leave behind M^- (d^2) in the valence state configuration $d(z^2)^2$.

From Griffith's tables[21], we find that this is a linear combination of the terms $^1A_{1g}(e_g^2)$ and $^1Eg(e_g^2)$; thus at large $M^- - R^+$ separations the valence state would correlate with the more stable of these, which is ^1E_g. The next step is to consider the terms of the ground state configuration t_{2g}^2. As it happens, t_{2g}^2 also includes a term of type ^1E_g; thus the non-crossing rule comes into operation. The extent to which the curves each avoid the crossing-point is given by the off-diagonal matrix element between them. This must be restricted to the two-electron part of the Hamiltonian, since the configurations differ by more than one electron, and is thus (nephelauxetic effects aside) independent of the environment. From Griffith's tables and the form of the valence state, the relevant matrix element is between $\sqrt{6}$ B and $2\sqrt{3}$ B depending on the position of the attempted crossover. B is a Racah parameter which for a first row transition metal is of the order of $800-1000$ cm^{-1}. Thus as Fig. 6 shows, the potential energy curve at crossover is lowered by some $1500-3000$ cm^{-1}. This quite considerable effect occurs in a region of incomplete dissociation of the metal-carbon bond (Fig. 6) and so might be expected to affect the interaction energy. In addition, the configuration interaction described here ensures that the valence pair electrons of M$^-$ are both smoothly transferred from e_g to t_{2g} orbitals. To summarise, R$^-$ loss from a d^0 system is not specially assisted; R$\cdot$ loss can be assisted by the demotion of one electron provided the system achieves a symmetry crossover; and R$^+$ loss is facilitated both by the demotion of two electrons and by the term-term interaction that ensures that demotion.

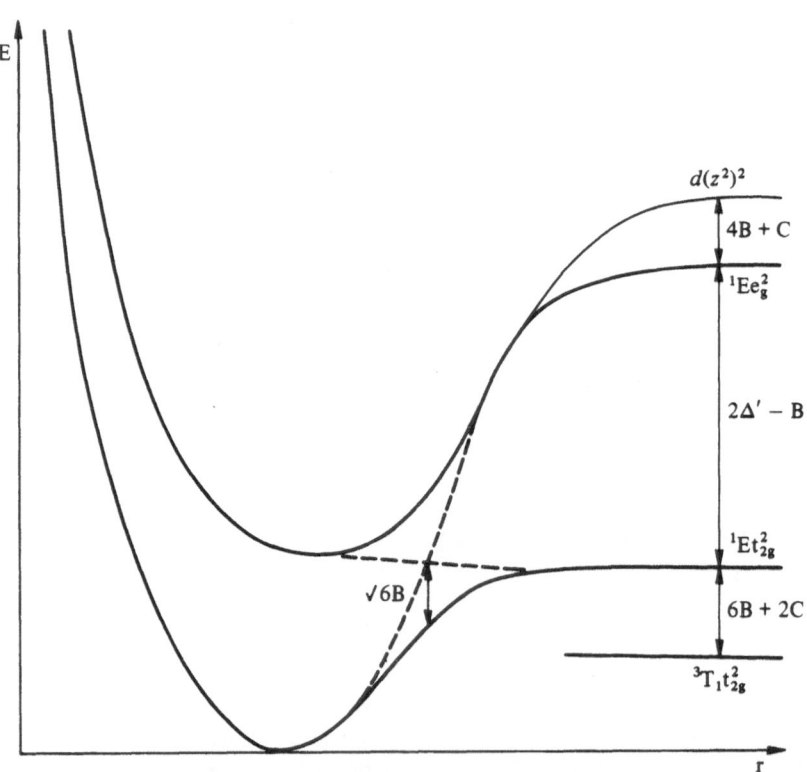

Fig. 6. Formal R$^+$ loss from an octahedral d^0 metal

Table 2. Configurational effects in ligand loss from low-spin octahedral species

	R^- loss	$R\cdot$ loss	R^+ loss[a]
d^0	$-$[b]	1, unassisted[c]	2, *assisted*
d^1	$-$	1, *assisted*	2, *assisted*
d^2	$-$	1, *assisted*	2, *assisted*
d^3	$-$	1, unassisted	1, unassisted
d^4	$-$	1, unassisted	1, *assisted*
d^5	$-$	1, unassisted	1, unassisted
$d^6 - d^{10}$	$-$	$-$	$-$

[a] See text for use of these expressions
[b] Blank signifies no first order effects
[c] I. e. process favoured by demotion of one electron; but not assisted by configuration interaction

The extension to other cases is straightforward but tedious, and the principal results for low-spin octahedral species are summarised in Table 2, which shows some interesting features. At this level of discussion, R^- loss is never assisted. The question of demotion only arises where the t_{2g} subshell is less than full. Two-electron demotion is is only possible for R^+ loss from d^0, d^1, and d^2 systems, and in all of these it is actually term-term assisted. R^+ loss is assisted by the demotion of one electron for d^3, d^4, or d^5 curves, but among these it is only term-term assisted for d^4. $R\cdot$ loss is clearly assisted by the demotion of a single electron for all d^n (n < 6), but is only term-term assisted for n = 1 and n = 2. (These predictions are quite different from those of Ref.[63], which refers exclusively to second order terms in R^- loss). The only configurations with n < 6 for which *no* process shows first order term-term assistance are d^3 and d^5. This is a gratifying result and tends to promote confidence in the usefulness of the theory. The relative ease of preparation of Cr(III) alkyl complexes has often been noted[71], and d^5 is exemplified by the Co(IV) alkyls now known to be accessible by electrochemical oxidation of Co(III)[72]. Presumably a parallel chemistry of Fe(III) awaits discovery.

C. Other Geometries

The extension to non-octahedral complexes is possible, but must be carried out with great care[73]. The orbitals chosen can be expressed as a linear combination of the usual orbitals for an octahedron, and electron-electron repulsions can then be calculated from those[21] for the octahedral case. It is not necessarily adequate for tetrahedral complexes of the first row transition elements, to use ligand field theory in the strong field limit, even for powerful ligands; in $V(mesityl)_4^-$ the ligand field splitting is only 9250 cm^{-1} [74].

V Conclusions and Implications

Oxidative or reductive coupling of organic ligands attached to a transition metal can often be treated formally as electrocyclic processes. The treatment leads to a simple extension of the Woodward-Hoffmann rules for such reactions. A number of caveats remain, however.

Energy gaps within an incompletely filled d subshell are smaller than the gaps usually encountered in purely organic systems. Thus the degree of selectivity may be smaller. In such cases, additional ligands may increase ligand field splittings and, hence, selectivities.

Transformations of organic groupings under the influence of a metal are usually the result of a sequence of steps each one of which must be examined. The plausibility of each proposed step depends on the plausibility of the overall chemistry at the metal, as well as on orbital energies and densities within the actual metal-organic grouping. Thus other ligands may exert profound effects.

There may be changes in orbital phase relationships, and in reaction energetics, as a result of geometrical flexibility in the coordination sphere of the metal. Reduction of flexibility by chelation may then profoundly affect reactivity.

There may be special difficulties in reactions where the ordering of orbitals centred on the metal changes along the actual reaction path, because of configuration interaction and the non-crossing rule for states.

Processes involving only one metal-carbon bond are even more subtle. Ligand loss as R^- avoids formal reduction of the metal and shows no new features, but where formal reduction does occur the fate of the electrons initially "in" the metal-carbon bond will depend on the properties of the d^n manifold produced. Orbital correlation theory becomes inadequate, but the results of state correlation theory are encouraging, with the low lability of Cr(III) and Co(IV) alkyls being correctly predicted.

There remains the intractable problem of alternative mechanisms. If theory predicts that some species should react readily by a specified route, and this happens, the theory is to that extent confirmed. If it fails to react the theory is wanting and must be modified. But if the species reacts readily by some pathway outside the scope of the theory, the theory remains untested and becomes irrelevant. It is thus a matter of some importance to be able to predict the influence of the d^n manifold on many-bond processes such as α- and β-elimination and reductive elimination. We might try to do this by analogy, but the analogies themselves are questionable. For instance, we may predict that internal abstraction is related to $R\cdot$ loss, since both involve an essentially electroneutral M-R bond cleavage. This ignores the changed effect on the metal of the hydrogen-donating ligand.

There is an obvious and urgent need for more systematic kinetic evidence and isotope effect studies, especially for the less straightforward processes, and even purely qualitative observations of relative reactivities, solvent and spectator ligand effects, selectivity, and the nature of minor products would be of the greatest value.

Acknowledgements. It is a pleasure to thank the Gesellschaft Deutscher Chemiker for asking me to give the lecture[1] on which this article is based. It is also a pleasure to thank Professor R. H. Grubbs, Dr. J. Carnduff, Dr. R. J. Cross, Professor P. Heimbach, Dr. B. Webster, and Professor A. Yamamoto for helpful and cordial discussions and correspondence. This is not to imply agreement; concensus chemistry remains mercifully remote.

VI References

1. Based in part on a lecture to the 2nd FECHEM Conference on Organometallic Chemistry, Hameln, September 1978
2. Braterman, P. S., Cross, R. J.: Chem. Soc. Rev. *2*, 271 (1973)
3. Mango, F. D., Schachtschneider, J. H.: J. Am. Chem. Soc. *89*, 2484 (1967)
4. Noyori, R., Yamakawa, M., Takaya, H.: J. Am. Chem. Soc. *98*, 1470 (1976)
5. See e. g. Fukui, K.: Theory of orientation and stereoselection. Berlin, Heidelberg, New York: Springer 1975
6. Woodward, R. B., Hoffmann, R.: The conservation of orbital symmetry. New York: Academic Press 1969
7. Davies, S. G., Green, M. L. H., Mingos, D. M. P.: Nouveau, Chimie *1*, 445 (1978)
8. Batich, C. D.: J. Am. Chem. Soc. *98*, 7585 (1976)
9. Pearson, R. G.: Symmetry rules for chemical processes. New York: Wiley-Interscience 1976
10. Yamamoto, T., Yamamoto, A., Ikeda, S.: J. Am. Chem. Soc. *93*, 3350 (1971)
11. Yamamoto, A.: Private communication
12. Chatt, J., Shaw, B. L.: J. Chem. Soc. *1959*, 705
13. Braterman, P. S., Cross, R. J., Young, G. B.: J. C. S. Dalton Trans. *1976*, 1306
14. Braterman, P. S., Cross, R. J., Young, G. B.: J. C. S. Dalton Trans. *1976*, 1310
15. Braterman, P. S., Cross, R. J., Young, G. B.: J. C. S. Dalton, Trans. *1977*, 1892
16. Komiya, S., Albright, T. A., Hoffmann, R., Kochi, J. K.: J. Am. Chem. Soc. *98*, 7255 (1976)
17. Komiya, S., Albright, T. A., Hoffmann, R., Kochi, J. K.: J. Am. Chem. Soc. *99*, 8440 (1977)
18. Morrell, D. G., Kochi, J. K.: J. Am. Chem. Soc. *97*, 7262 (1975)
19. Young, G. B., Whitesides, G. M.: J. Am. Chem. Soc. *100*, 5805 (1978)
20. Braterman, P. S.: J. C. S. Chem. Comm. *1979*, 70
21. Griffith, J. S.: The theory of transition metal ions. Cambridge: CUP 1961
22. Mango, F. D., Schachtschneider, J. H.: J. Am. Chem. Soc. *93*, 1123 (1971)
23. Braterman, P. S., Cross, R. J.: J. C. S. Dalton Trans. *1972*, 657
24. Dyachkovsky, F. S., Khrushch, N. E.: Zh. Obshch. Khim. USSR *41*, 1779 (1971); compare Khachaturov, A. S., Bresler, L. S., Poddubnyi, I. Ya.: J. Organometal. Chem. *42*, C18 (1972)
25. Occam, Wm. of, attrib.
26. Grandjean, J., Laszlo, P., Stockis, A.: J. Am. Chem. Soc. *96*, 1623 (1974)
27. Grevels, F.-W., Schultz, D., Koerner von Gustorf, E. A.: Angew. Chem. Intern. Edn. *13*, 534 (1974)
28. Koerner von Gustorf, E. A.: Chemical Society Centenary Lectures 1975; cf Kerber, R. C., Koerner von Gustorf, E. A.: J. Organometal. Chem. *110*, 345 (1976)
29. Houk, K. N.: J. Am. Chem. Soc. *95*, 4092 (1973)
30. Pearson, R. G.: Topics in Curr. Chem. *41*, 75 (1973)
31. See e. g. Beach, N. A., Gray, H. B.: J. Am. Chem. Soc. *90*, 5713 (1968)
32. Mahnke, H., Sheline, R. K.: Inorg. Chem. *15*, 1245 (1976)
33. Grubbs, R. H., Miyashita, A., Liu, M.-I. M., Burk, P. L.: J. Am. Chem. Soc. *99*, 3863 (1977)
34. Grubbs, R. H., Miyashita, A.: J. Am. Chem. Soc. *100*, 1300 (1978)
35. McDermott, J. X., Wilson, M. E., Whitesides, G. M.: J. Am. Chem. Soc. *98*, 6529 (1976)
36. Krausse, J., Schlödl, G.: J. Organometal. Chem. *27*, 59 (1971)
37. Perkins, D. L., Puddephatt, R. J., Tipper, C. F. H.: J. Organmetal. Chem. *154*, C16 (1978)
38. Grubbs, R. H.: J. Organometal. Chem. Library *1*, 423 (1976)

39. Fenske, R. F.: Prog. Inorg. Chem. *21*, 179 (1976)
40. Wilke, G., Heimbach, P.: Angew. Chem. *75*, 10 (1963)
41. Buchholz, H., Heimbach, P., Hey, H. J., Selbeck, H., Wiese, W.: Coord. Chem. Rev. *8*, 129 (1972)
42. Heimbach, P., et al.: to be published
43. Cassar, L., Eaton, P. E., Halpern, J.: J. Am. Chem. Soc. *92*, 3515 (1970)
44. Dauben, W. G., Kielbania, A. J., Jr.: J. Am. Chem. Soc. *93*, 7345 (1971)
45. Hogeveen, H., Volger, H. C.: J. Am. Chem. Soc. *89*, 2486 (1967)
46. Cassar, L., Eaton, P. C., Halpern, J.: J. Am. Chem. Soc. *92*, 6366 (1970)
47. Paquette, L. A., Stowell, J. C.: J. Am. Chem. Soc. *92*, 2584 (1970); Paquette, L. A., Beckley, R. S., McCreadie, T.: Tetrahedron Letters *1971*, 775
48. Dauben, W. G., Buzzolini, M. G., Schallhorn, C. H., Whalen, D. L.: Tetrahedron Letters *1970*, 787; Dauben, W. G., Schallhorn, C. H., Whalen, D. L.: J. Am. Chem. Soc. *93*, 1446 (1971)
49. Paquette, L. A., Accounts Chem. Res. *4*, 280 (1971)
50. Mango, F. D.: Topics in Curr. Chem. *45*, 39 (1974)
51. Weigert, F. J., Baird, R. L., Shapley, J. R.: J. Am. Chem. Soc. *92*, 6630 (1970)
52. Volger, H. C., Hogeveen, H. C.: Rec. Trav. Chim. Pays-Bas *86*, 830 (1967)
53. Volger, H. C., Gaasbeek, M. M. P.: Rec. Trav. Chim. Pays-Bas *87*, 1290 (1968)
54. Slegeir, W., Case, R., McKennis, J. S., Pettit, R.: J. Am. Chem. Soc. *96*, 287 (1974)
55. See e. g. Maitlis, P. M.: Pure App. Chem. *33*, 489 (1973)
56. Mango, F. D., Schachtschneider, J. H.: J. Am. Chem. Soc. *91*, 1030 (1969)
57. Yamazaki, H., Wakatsuki, Y.: J. Organometal. Chem. *149*, 377 (1978), and references therein
58. Yamazaki, H., Wakatsuki, Y.: J. Organometal. Chem. *139*, 157 (1977)
59. Wakatsuki, Y., Yamazuki, H.: J. Organometal. Chem. *139*, 169 (1977)
60. Wakatsuki, Y., Yamazuki, H.: J. Organometal. Chem. *149*, 385 (1978)
61. Cossee, P.: J. Catalysis *3*, 80 (1964)
62. Baird, M. C.: J. Organometal. Chem. *64*, 289 (1974)
63. Mingos, D. M. P.: J. C. S. Chem. Somm. *1972*, 165
64. Braterman, P. S.: J. C. S. Chem. Comm. *1972*, 761
65. Braterman, P. S.: unpublished results
66. Dodd, D., Johnson, M. D., Lockman, B. L.: J. Am. Chem. Soc. *99*, 3664 (1977)
67. Yamamoto, A.: 176th ACS National Meeting, Miami, September 1978, Abstract INOR 109
68. Murrell, J. N.: J. C. S. Chem. Comm. *1972*, 1044
69. Dewar, M. J. S.: Chemistry in Britain *11*, 97 (1975)
70. Schaffer, C. Jorgensen, C. K.: Mol. Phys. *9*, 401 (1965)
71. Sneeden, R. P. A.: Organochromium compounds. London, New York: Academic Press 1975
72. Levitin, I., Sigan, A., Vol'pin, M. E.: J. C. S. Chem. Comm. *1975*, 469
73. Braterman, P. S.: unpublished
74. Seidel, W., Kreisel, G.: Zeit. Anorg. Allg. Chem. *426*, 150 (1976)

Received April 9, 1979

Author Index Volumes 50–92

Renger, G.: Inorganic Metabolic Gas Exchange in Biochemistry. *69*, 39–90 (1977).

Rice, S. A.: Conjuectures on the Structure of Amorphous Solid and Liquid Water. *60*, 109–200 (1975).

Rieke, R. D.: Use of Activated Metals in Organic and Organometallic Synthesis. *59*, 1–31 (1975).

Rodehorst, R. M., see Koch, T. H.: *75*, 65–95 (1978).

Roychowdhury, U. K., see Venugopalan, M.: *90*, 1–57 (1980).

Rüchardt, C.: Steric Effects in Free Radical Chemistry. *88*, 1–32 (1980).

Ruge, B., see Dürr, H.: *66*, 53–87 (1976).

Sandorfy, C.: Electronic Absorption Spectra of Organic Molecules: Valence-Shell and Rydberg Transitions. *86*, 91–138 (1979).

Sargent, M. V., and Cresp, T. M.: The Higher Annulenones. *57*, 111–143 (1975).

Schacht, E.: Hypolipidaemic Aryloxyacetic Acids. *72*, 99–123 (1977).

Schäfer, F. P.: Organic Dyes in Laser Technology. *68*, 103–148 (1976).

Schenkluhn, H., see Heimbach, P.: *92*, 45–107 (1980).

Schneider, H.: Ion Solvation in Mixed Solvents. *68*, 103–148 (1976).

Schnoes, H. K., see DeLuca, H. F.: *83*, 1–65 (1979).

Schönborn, M., see Jahnke, H.: *61*, 133–181 (1976).

Schuda, P. F.: Aflatoxin Chemistry and Syntheses. *91*, 75–111 (1980).

Schuster, P., Jakubetz, W., and Marius, W.: Molecular Models for the Solvation of Small Ions and Polar Molecules. *60*, 1–107 (1975).

Schwarz, H.: Some Newer Aspects of Mass Spectrometric *Ortho* Effects. *73*, 231–263 (1978).

Schwedt, G.: Chromatography in Inorganic Trace Analysis. *85*, 159–212 (1979).

Sears, P. G., see Lemire, R. J.: *74*, 45–91 (1978).

Shaik, S., see Epiotis, N. D.: *70*, 1–242 (1977).

Sheldrick, W. S.: Stereochemistry of Penta- and Hexacoordinate Phosphorus Derivatives. *73*, 1–48 (1978).

Simonis, A.-M., see Ariëns, E. J.: *52*, 1–61 (1974).

Simons, J. P., see Ashfold, M. N. R.: *86*, 1–90 (1979).

Sluski, R. J., see Koch, T. H.: *75*, 65–95 (1978).

Smith, D., and Adams, N. G.: Elementary Plasma Reactions of Environmental Interest, *89*, 1–43 (1980)

Sørensen, G. O.: New Approach to the Hamiltonian of Nonrigid Molecules. *82*, 97–175 (1979).

Spanget-Larsen, J., see Gleiter, R.: *86*, 139–195 (1979).

Špirko, V., see Papoušek, D.: *68*, 59–102 (1976).

Stuhl, O., see Birkofer, L.: *88*, 33–88 (1980).

Sutter, D. H., and Flygare, W. H.: The Molecular Zeeman Effect. *63*, 89–196 (1976).

Tacke, R., and Wannagat, U.: Syntheses and Properties of Bioactive Organo-Silicon Compounds. *84*, 1–75 (1979).

Tsigdinos, G. A.: Heteropoly Compounds of Molybdenum and Tungsten. *76*, 1–64 (1978).

Tsigdinos, G. A.: Sulfur Compounds of Molybdenum and Tungsten. Their Preparation, Structure, and Properties. *76*, 65–105 (1978).

Tsuji, J.: Applications of Palladium-Catalyzed or Promoted Reactions to Natural Product Syntheses. *91*, 29–74 (1980).

Ugi, I., see Dugundji, J.: *75*, 165–180 (1978).

Ullrich, V.: Cytochrome P450 and Biological Hydroxylation Reactions. *83*, 67–104 (1979).

Venugopalan, M., Roychowdhury, U. K., Chan, K., and Pool, M. L.: Plasma Chemistry of Fossil Fuels. *90*, 1–57 (1980)

Vepřek, S.: A Theoretical Approach to Heterogeneous Reactions in Non-Isothermal Low Pressure Plasma. *56*, 139–159 (1975).

Vepřek, S., see Gruen, D. M.: *89*, 45–105 (1980)

Reactivity and Structure

Concepts in Organic Chemistry

Editors: K. Hafner, J.-M. Lehn, C. W. Rees,
P. v. Ragué Schleyer, B. M. Trost,
R. Zahradnik

This series will not only deal with problems
of the reactivity and structure of organic
compounds but also consider synthetical-
preparative aspects.
Suggestions as to topics will always be
welcome.

**Springer-Verlag
Berlin Heidelberg New York**

Polymers Properties and Applications

Editorial Board: H.-J. Cantow,
H. J. Harwood, J. P. Kennedy,
A. Ledwith, J. Meißner,
S. Okamura, G. Olivé,
S. Olivé

Volume 3: A. Knop, W. Scheib

Chemistry and Application of Phenolic Resins

1979. 111 figures, 88 tables. XIII, 269 pages
ISBN 3-540-09051-7

The authors present the current theory of phenolic resin chemistry and the technical application of phenolic resins, based on day-to-day experience in research, production and marketing, and against the background of economic relevance. Where the first fully synthetic polymers (phenolic resins) stand today and what their future is are subjects of discussion. Looking back at their development, it is shown that after a wide variety of adaptions, they remain technically and economically irreplaceable products with potential for further market growth and a commensurate appreciation of their value. This book will be greatly appreciated by chemists, engineers, marketing professionals, and students.

Volume 2: H.-H. Kausch

Polymer Fracture

1978. 180 figures, 23 tables. X, 332 pages
ISBN 3-540-08786-9

"Kausch,... is well known for his work on polymer morphology and molecular mechanics as well as his research on the strength of materials. The avowed aim of this book is to connect the more conventional statistical and continuum mechanics interpretation of fracture phenomena to the newer spectroscopic studies of highly stressed polymeric chains and the kinetics of their rupture. Relating the literature on the observed modes of viscoelasticity and irreversible deformation from polymer morphology and solid-state physics, Kausch explains the behavior and rupture of polymeric materials in terms of molecular slip and breakage processes. This leads to interesting, methodical and well-thought-out interpretations of fracture toughness, crack propagation rates and fatigue of all major polymer systems... Thus, the book is an outstanding contribution to our understanding of the role of chain ruptures during mechanical failure... every student and practitioner of polymer science and engineering should find this book to be a valuable resource for his work."
Physics Today

Springer-Verlag
Berlin
Heidelberg
New York

Volume 1: B. Rånby, J. F. Rabek

ESR Spectroscopy in Polymer Research

1977. 356 figures, 29 tables. XIV, 410 pages
ISBN 3-540-08151-8

"...This book is a remarkable example for the successful combination of simplicity and clarity in its tutorial parts and of depth and width whenever and wherever is presents the state of the art...As ultimate and very gratifying reward for his investment the reader gets no less than 2519 references to the literature in excellent alphabetical order. Scientists who already work with ESR will be greatly assisted in their efforts by this book; those who do not yet use this method will have an easy time to learn and use it. All of them will be grateful to the authors for this exceptional addition to our scientific literature."
J. Polymer Science